AUTOWAVE PLASTICITY

AUTOWAVE PLASTICITY

LOCALIZATION AND COLLECTIVE MODES

Lev Zuev

CRC Press is an imprint of the
Taylor & Francis Group, an **informa** business

Translated from Russian by V.E. Riecansky

CRC Press
Taylor & Francis Group
6000 Broken Sound Parkway NW, Suite 300
Boca Raton, FL 33487-2742

Printed on acid-free paper

International Standard Book Number-13: 978-0-367-85681-6 (Hardback)

Visit the Taylor & Francis Web site at
http://www.taylorandfrancis.com

and the CRC Press Web site at
http://www.crcpress.com

Contents

Preface

In the science of plasticity there are two ways – scientific and engineering. The researchers who chose the first of them, starting from a theoretical description of the structure of ideal and real solids, are moving towards an explanation of the phenomenon of plastic deformation in terms of physics, mechanics and chemistry of a solid body. The criterion of success here often has an aesthetic sense. Proponents of the engineering path start with the problems of technology of manufacturing processes of products by deforming materials, and are mainly interested in the behavior of materials during such processes, as well as in extreme conditions of operation. A sign of success on this path is the usefulness of the results obtained for practice.

However, the goal pursued by scientists and engineers, who sometimes follow these different paths independently of one another, is the same — an in-depth and comprehensive understanding of the nature of plasticity, capable of ensuring progress both in theory and in improving modern industrial technologies. For this reason, the result achieved when moving along one of the paths will undoubtedly be claimed by supporters of the other.

This book is mainly devoted to the scientific aspect of the problem of plasticity. It develops and interprets the experimental results partially described earlier in our monograph, and also obtained in recent years by the author and his colleagues in the Laboratory of Strength Physics of the Institute of Strength Physics and Materials Science, Siberian Division of the Russian Academy of Science (Tomsk).

The time elapsed since the publication of this book was given to attempts to create a consistent system of views on the process of plastic flow, which is developing at a macroscopic scale level. The main goal during these years was to introduce, experimentally verify and logically present such a system. Within its framework, plastic deformation is characterized by localization patterns that regularly develop during the entire process of deformation, and is considered as a process of structure formation in a deformable medium. Such processes constitute the content of synergetics, an interdisciplinary field of research that has been booming in recent decades.

The results obtained during this time, and their generalization in this monograph led to the formation of a new plasticity model based on the idea of localization of plastic flow, and made it possible to understand its main role in the deformation process of solids.

The present monograph consists of five chapters.

Chapter 1 is devoted to the current state of research and the rationale for choosing the most promising, according to the author, areas of analysis of the phenomenon of plasticity. The huge amount of work on these issues makes it impossible to present here any detailed literature review. Therefore, in Chapter 1, attention is focused mainly on those problems of plastic flow which usually cause a dead end for researchers, and on the search for a way out of these situa*tions*.

Chapter 2 presents experimental material on the nature of macrolocalization of plastic flow at all stages of the plastic flow process. We describe the method of observing localized plasticity using speckle photographs used in our work, consider the types of observed patterns of localized plasticity, and systematize them. A one-to-one correspondence was established between the patterns and the stages of strain hardening.

Chapter 3 contains an analysis of the possibility of an autowave description of macrolocalization of the course. It shows that the patterns of localized plasticity can be considered as different modes of autowave processes. The mode is uniquely determined by the law of strain hardening, acting at the appropriate stage of the process, and changes when the mechanism of strain hardening changes.

Chapter 4 introduces and discusses a two-component model for the development of localized plasticity and formulates the concept of an elastoplastic deformation invariant, which serves as the mathematical expression of the proposed model. It has been shown that many important regularities in the development of localized plasticity are consequences of the elastoplastic deformation invariant.

Chapter 5 describes an attempt to use a quasiparticle approach in an autowave model of plastic flow. Here, in the framework of the accepted physics of the condensed state of the method, a quasi-particle is introduced that corresponds to the autowave of localized plasticity and called autolocalizon, its characteristics are evaluated, and its use to describe plastic flow processes is considered. The relationship between the plasticity of metals and the position of the elements in the Periodic Table of Elements is also considered.

The author is grateful to the staff of the Laboratory of Physics of Strength of the Institute of Strength Physics and Materials Science of the Siberian Branch of the Russian Academy of Sciences, with whom all the experimental results were obtained.

The scientific aspects of the developed approach were discussed with professors Yu.A. Khon, S.A. Barannikova, V.I. Danilov, A.I. Olemskoi, V.I. Vettegren, V.I. Betekhtin, G.A. Malygin, A.M. Glezer, O.B. Naimark, V.E. Wildemann, Yu.I. Chumlyakov, Yu.A. Alyushin, E.M. Morozov, V.I. Alshits, V.M. Chernov, G.A. Malygin, P.V. Makarov, V.M. Zhigalkin, as well as E.E. Glikman (Israel), E. Aifantis, I. Karaman, V.M. Finkel (USA), I.A. Abrikosov (Sweden), V.Z. Bengus, V.V. Pustovalov (Ukraine). Thanks to all of them for their advice and useful criticism!

The Physical Seminar of the Institute of Strength Physics and Materials Science of the Siberian Branch of the Russian Academy of Sciences, to which I want to express my gratitude, played a huge role in developing a new system of representations.

Introduction

Describing the state of the plasticity science, Bell wrote in 1973: "In a strikingly large number of cases known to researchers, there is still no reliable knowledge corresponding to plausible theories that could explain the experimental observations already performed in this field" (see [Bell, 1984]).

And today, after more than half a century, we have to admit that the long history of research into the nature of plastic deformation of solids has not led to the creation of a generally accepted system of views on the nature of this phenomenon. The problem of the plasticity of solids turned out to be much more difficult than it could be imagined, and by now only the main reasons for the difficulties of its solution have been found. They are determined by the extraordinary complexity of the shape of the response of the deformable medium to the external influence; this complexity is illustrated by the diversity of the stress–strain curves for different materials and loading conditions.

The complexity of the problem of plasticity is associated with such properties of a deformable medium as its nonlinearity and activity, as well as the ability to remember the effects on it resulting from structural changes. Adequate and simultaneous consideration of all these properties in the experimental study and theoretical description of the plastic deformation of materials is extremely complex, both in conceptual and mathematical sense. In addition, the awareness of the need to take into account these properties has developed only relatively recently.

Attempts to construct microscopic models of plasticity in the last 60–70 years have been almost completely based on the results of using electronic transmission microscopy of dislocations and dislocation assemblies in deformed metals. Remarkable progress has been made in this area, but, unfortunately, a satisfactory physical theory of plasticity has not asbeen built.

This disappointing outcome was caused by two reasons. The first is related to the fundamental features of the technique of electron-microscopic analysis of thin foils, when the study is separated in time and space from the actual process of plastic flow.

The second reason is that the development of a traditional microscopic approach to the problem of plasticity was complicated by the continuous identification of a huge number of process details individual for different materials. Because of this, their simultaneous recording and coordinated interpretation became almost impossible. The range of difficulties that impede the solution of the problem has expanded as more and more high-resolution techniques were applied and as new research proceeded.

A paradoxical situation has arisen: the microscopic technique has made it possible for researchers to obtain rich data on the structure of deformable metals, but the compilation and understanding of these data turned out to be impossible. The end of the twentieth century in studies of plasticity was marked by an understanding of these circumstances and the appearance of the first studies that took into account the multiscale nature of the processes of plastic flow and the structure of a deformable medium [Panin, Likhachev, Grinyayev, 1985; Likhachev, Malinin, 1994]. In the last decades, the nonlinearity of the deformable medium, which was recognized as crucial for the description of plasticity, was taken into account, since it determines the nature of the most important deformation patterns [Aifantis, 1996].

At the same time, it became clear that the simplicity and clarity of model representations should be sought within the macroscopic approach, designed to describe the localization of plastic deformation [Zuev, 1994, 1996, 2006a, b; Zuev, Danilov, Gorbatenko, 1995]. Much less attention is paid to this side of the processes of plastic flow, which is largely due to the impossibility of observing the localization of plastic deformation. There was a need to create methods for the study of deformation, providing in situ observation of deformation processes at the macroscopic scale level. Such problems were solved in the last quarter of the twentieth century due to the use of various variants of holographic interferometric methods for measuring strain obtained in those years [West, 1982].

These ideas and their experimental verification became the basis of an autowave approach to the problem of the plasticity of solids

[Zuev, Danilov, Barannikova, 2008]. This approach touched the basics of the existing ideas about the problem of plasticity. It was found that the most important regularity of plastic flow is its localization, which is implemented during the whole process, acquiring different forms. Localization identifies the most interesting volumes of a deformable medium, in which the deformation overtakes in its development the processes proceeding in other volumes. It is in the regions of localization that the mechanisms of strain hardening and fracture are activated. The very existence of regions for localizing plastic deformation means that the deformation processes in the medium spontaneously form the structure, reducing the symmetry of the system [Wigner, 1971].

The fundamental difference between the concepts of structure in the theory of elasticity and the theory of plasticity should be emphasized. With elastic deformation, the structure is static, that is, the boundaries of the structural elements in a multiphase medium vary only slightly with deformation [Eringen, 1975; Shermergore, 1977]. On the contrary, a new structure of a deformable medium arises in the course of plastic flow [Seeger, Frank, 1988], that is, new defects of the crystal structure and their ensembles of different scale and complexity appear.

When discussing the results of research, a physicomechanical approach to the problem of plasticity was used. Ideas about the self-organization of deformable media were borrowed from physics, and the experimental mechanics of a deformable solid provided the necessary information about the features and laws of the plastic flow of materials of different nature. This approach determined the range of issues under consideration and the possibility of understanding them. First of all, the author was interested not in the mechanical structural properties of the material [Makhutov, 2005], but in the course of dynamic processes in an environment capable of self-organization during plastic flow [Seeger, Frank, 1987]. The object of study in this case is the continuum, capable of generating spatial-time periodic processes with different types of autowaves with uniform deformation.

In presenting the material, the author tried to bring explanations at least to a rough numerical estimate, trying to make the observed facts and considerations about their nature more understandable for themselves and for readers. The ‘roughness’ of the estimates is justified by the fact that, by resorting to them, the author followed the opinion of Ziman [1962], who wrote: "Complicated calculations often give a plausible look to some very dubious assumptions."

Working on a monograph, the author consciously limited himself to discussing only the most general laws of plastic flow, characteristic of all deformable media, regardless of their nature, structure, and deformation mechanism. The particular features of the deformation of concrete metals and alloys were deliberately ignored. Of course, autowaves of localized plasticity in copper differ from autowaves in iron, but the generality of the phenomenon of autowave deformation itself actualizes this approach. The established regularities are explained by attracting the ideas of synergetics to the problem of plastic flow, the analysis of micromechanisms of which is based on the theory of defects in the crystalline structure.

Concise terminological dictionary[1]

Autowave (*self-excited wave)* is a self-sustaining wave in an active medium, keeping its characteristics constant due to the energy source distributed in the medium [Krinsky, Zhabotinsky, 1981].

The active medium is characterized by the fact that: *a*) there is a distributed source of energy or substances rich in energy and negentropy; *b*) in the medium, one can distinguish an elementary volume of complete mixing, which contains an open point system, far from thermodynamic equilibrium; *c*) the connection between neighboring elementary volumes is due to the transfer phenomena "[Romanovsky, Stepanova and Chernavsky, 1984].

A quasiparticle is an elementary excitation of a condensed medium (solid, liquid), which in some respects behaves like a quantum particle. Such excitation is associated, as a rule, not with the movement of a single particle, but with the coordinated movement of many (or all) particles of the system [Kaganov, Lifshits, 1989].

Stress concentration – an increase in local stresses near abrupt changes in the shape of the surface of the body or near the places of action of external forces dramatically changing along the coordinates [Sedov, 1970. Vol. 2].

Plastic deformation localization is a phenomenon in which some specific part of the sample is plastically deformed, depending on the shape of the stress–strain diagram and the magnitude of the elastic modulus [McClintock, Argon. 1970].

Nonlinearity is a property of the medium, which means the multivariance of the paths of evolution, the presence of a choice from

[1] An explanation of the terms is given in the sense as they are used in the text of the book.

alternative paths and a certain rate of evolution, and the irreversibility of evolutionary processes. Nonlinearity in the mathematical sense means a certain kind of mathematical equations containing the desired quantities in powers greater than one, or coefficients depending on the properties of the medium. Nonlinear equations can have several (more than one) qualitatively different solutions. This implies the physical meaning of nonlinearity. The set of ways for the evolution of the system described by these equations corresponds to the set of solutions of a nonlinear equation [Knyazeva, Kurdyumov, 1994].

A **pattern** is any sequence of phenomena in time or any arrangement of objects in space that can be distinguished from another sequence or another arrangement, or compare them [Walter, 1966].

Self-organization is the acquisition of a spatial, temporal, or functional structure by a system without specific external influence [Haken, 2014]. The concept of self-organization describes processes that are far from equilibrium, which, through the driving forces acting in the system, lead to complex structures [Ebeling, 1979].

Synergetics is an interdisciplinary scientific field that studies the principles governing the emergence of self-organizing structures and (or) functions in systems of a different nature that are far from equilibrium [Haken, 1985].

Dissipative structure is an ordered configuration that appears outside the stability domain of the thermodynamic branch [Nikolis, Prigogine, 1979].

The theory of dislocations is a branch of solid-state physics in which the relationship between the plastic properties of crystals and the atomic structure is considered. In the theory of dislocations, a more realistic picture of the crystal structure is considered than in most other areas of solid state physics, since all lattice defects are fully taken into account [Cottrell, 1969].

Stability of the system – *classical stability*: the ability to maintain behaviour when the environment changes; – *structural stability*: the ability to maintain behaviour when the structure of the system itself changes, caused by the influence of the external environment [Casti, 1982].

1

Plastic flow
Important regularities

The problem of the physical nature of the plastic deformation of solids in modern science is paradoxical. Studies with nearly two centuries of history (see, for example, [Lüders, 1860; Larmor, 1892]) did not lead to a complete understanding of the nature of this phenomenon. For comparison, I note that the publication of Planck's work, It was not more than thirty years before the works of Heisenberg and Schrödinger who created the modern edition of quantum mechanics (Hund, 1980). The phenomenon of plastic deformation of solids turned out to be complex, and the reason for this complexity is caused by a combination of three factors.

First, the crystal lattice itself is complex and its irreversible deformation is the content of the problem of plasticity. This is true even in relation to ideal defect-free crystals and it turns out when meeting the books of Zeitz [1949], Peierls [1956], Kittel [1967, 1978], Zeiman [1962, 1966], Frenkel [1972, 1975], Kosevich [1972], Ashcroft and Mermin [1978] and many other authors.

Secondly, beginning in the 30s of the 20th century, viable theories of plastic flow associate its laws with the birth, motion, and interaction of lattice defects — dislocations [Reed, 1957; Cottrell, 1958; Indenbom, 1960, 1979; Seeger, 1960; Van Buren, 1962; Friedel, 1967; Hirth, Lothe, 1972; Kosevich, 1978; Smirnov, 1981; Mughrabi, 1983, 2001, 2004; Kuhlmann-Wilsdorf, 2002]. Dislocations are characterized by the Burgers vector $b \approx 10^{-10}$ m, and the formation of a macroscopic neck during the transition to viscous fracture [Ekobori, 1971; Shiratori, Miyoshi, Matsushita, 1986; Makhutov, 2005] covers the sample scale $l\_$ 10^{-2} m, that is,

$L/b \approx 10^8$. As a result, the difficulty arises of reconciling these scales, which is necessary for understanding the nature of plastic flow.

Thirdly, the huge variety of features of the plastic flow identified in the study of specific materials makes it difficult to isolate and analyze the general patterns necessary and sufficient to create an adequate theory of this complex phenomenon.

1.1. Multi-scale plastic flow heterogeneity

One of the key problems in the description of plasticity is the spatial-temporal heterogeneity of this phenomenon [Shtremel', 1997, 1999]. Spatial heterogeneity is known as the localization of plastic deformation, that is, its inhomogeneous distribution in the volume of a deformable body [Aifantis, 1995; Zbib, delaRubia, 2002; Ohashi, Kawamukai, Zbib, 2007]. This means that with plastic flow in a medium areas arise in which the kinetics of plastic deformation is different. In turn, the temporal part of the inhomogeneity looks like a non-monotonic dependence of the deforming stress on the deformation s(e) and is known as a jump-like deformation called the Portevin–Le Chatelier effect [1994; Krempl, 2001; Maugin, 2013; Shibkov et al., 2016].

Analysis of data on the processes of plastic flow indicates the existence of dislocation, mesoscopic and macroscopic spatial scales of deformation [Panin, Likhachev, Grinyaev, 1985; Vladimirov, 1987; Likhachev, Malinin, 1993; Zbib, de la Rubia, 2002; Huvier, 2009]. Let us consider the spatial-time regularities of plastic flow characteristic of these scales sequentially.

1.1.1. Dislocation level of deformation

For about a hundred years, the development of the plasticity science has been mainly connected with the application of the theory of dislocations to this problem. There is a rich monographic literature [Reed, 1957; Cottrell, 1958; Friedel, 1967; Van Buren, 1962; Eshelby, 1963; Kunin, 1965; Hirth and Lothe, 1972; Indenbom, 1979], which examines various aspects of the physics of crystal defects. The dominant trend in the development of dislocation studies consists in the specification of data on the spatial distribution of linear defects responsible for plastic deformation and their ensembles in deformed crystals [Amelinks, 1968; Hirsch et al., 1968; Messerschmidt, 2010]. Deformation-dependent patterns of dislocation distributions became the basis for most of the proposed strain hardening models, so that mostly microscopic approaches developed and survived in plasticity physics, and progress was determined mainly by increasing the resolution of microscopes and expanding the range of materials under study.

Dislocation models of plastic flow in most cases are based on the use of the Taylor–Orowan equation of dislocation plasticity [Oding, 1959; Orlov, 1983; Suzuki, Yoshinaga, Takeuchi, 1989]

$$d\varepsilon / dt \equiv \dot{\varepsilon} = \alpha b \rho_m V_{\mathrm{disl}}, \tag{1.1}$$

Equation (1.1) relates the strain rate $\dot{\varepsilon}$ to the microscopic characteristics of the dislocation structure: the Burgers vector of dislocations b, the density of mobile dislocations r_m, and the speed of their movement V_{disl} at stress σ (α is a numerical factor of geometric origin). For a long time it seemed (see, for example, the review of Nadgorny [1972]) that after finding the exact form of the functions $V_{\mathrm{disl}}(\sigma, T...)$ and $r(\varepsilon, T...)$, the description of the dislocation deformation mechanisms becomes a simple task.

With the first of these functions $V_{\mathrm{disl}}(\sigma, T...)$ the situation is quite satisfactory. Numerous experimental data compiled by Nadgorny [1972], Lubents [1974] and Suzuki, Yoshinaga and Takeuchi [1989], as well as a theoretical analysis of dislocation mobility developed by Indenbom, Orlov and Estrin [1972], Engelke, [1973], Alshitz, Indenbom [1975a, b], Silard and Martin [Caillard, Martin, 2003] showed that two modes of motion of dislocations can be realized:

– thermally activated (low stress) with speed

$$V_{\mathrm{disl}} = V_t \exp\left[-\frac{G(\sigma)}{k_B T}\right] = V_t \exp\frac{\Delta S}{k_B} \exp\left(-\frac{U - \gamma\sigma_{ef}}{k_B T}\right) \sim e^{\sigma_{ef}}, \tag{1.2}$$

– quasi-viscous (high stress) with speed

$$V_{disl} = \frac{b\sigma_{ef}}{B(T)} \sim \sigma_{eff}. \tag{1.3}$$

In the relations (1.2) and (1.3), G= $U - T\Delta S - \gamma\sigma_{ef}$ is the Gibbs thermodynamic potential [Kubo, 1970], U is the height of the potential barrier overcome by a dislocation during movement, ΔS is the change of entropy during the movement of a dislocation, γ is the activation volume of the process, B is the coefficient of viscous

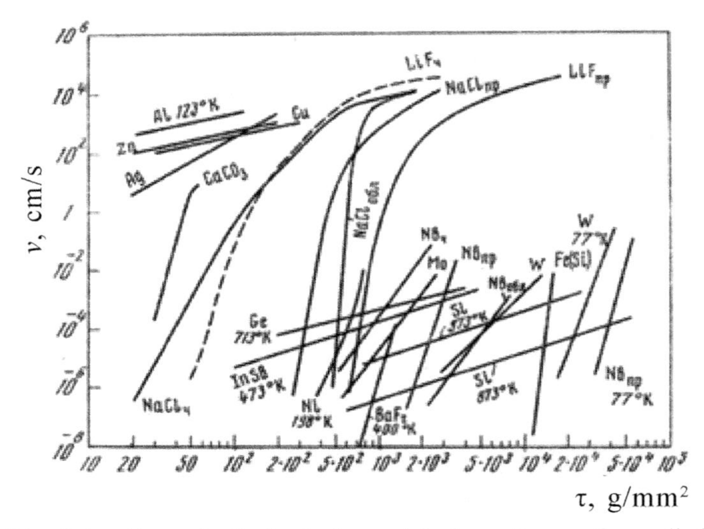

Fig. 1.1. The dislocation velocity in single crystals depending on the applied voltage at 300 K [Nadgorny, 1972]. (In Chapter 1, the designation of quantities in the figures is given in accordance with the quoted sources).

dragging of dislocations. A summary of experimental data on the mobility of individual dislocations in various crystals is shown in Fig. 1.1. The given data fully satisfy the equations (1.2) and (1.3).

The two modes of motion of dislocations are closely related to each other. The speed calculated by the formula (1.2) is effective: its evaluation takes into account the time it takes to stop dislocations at local barriers [Orlov, 1973, 1983; Suzuki, Yoshinaga, Takeuchi, 1983; Caillard, Martin, 2003]. Direct displacement of dislocations in a crystal between local barriers, as shown by Komnik and Bengus [1966], always occurs at a rate determined by the relation (1.3).

With the function $\rho_m(, \ldots T)$, also necessary for calculations by equation (1.1), the situation is much worse. When the total dislocation density $\rho_{tot} \geq 10^{12}$ cm^{-2} is reached at large degrees of deformation, the distance between them decreases to $\rho \approx \sigma\rho_{tot}^{-1/2} \leq 10^{-8}$ m, so that the dislocation individuality is lost, and the possibility of their observation by electron microscopy of thin foils is limited. The main thing is that the density of mobile dislocations during the process varies non-monotonously over a wide range [Reed, 1957; Cottrell, 1958; Wirtman, 1987; Hennecke, Hähner, 2009]. This question was considered in detail by Gilman [Gilman, 1968; Gilman, 1972], who showed that the density of mobile dislocations depends on the total strain ε_{tot} as

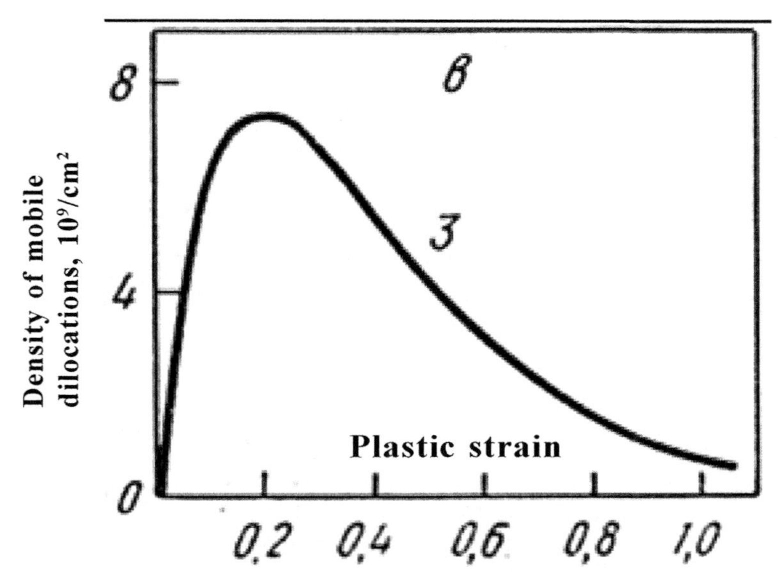

Fig. 1.2. The density of mobile dislocations depending on plastic deformation [Gilman, 1972].

$$\rho_m(\varepsilon) = (\rho_0 + M\varepsilon_{tot})\exp(-\Phi\varepsilon_{tot}), \tag{1.4}$$

where $M = 2m/b$, m is the multiplication factor of dislocations, $\Phi = \theta/\sigma$, $\theta = d\sigma/d\varepsilon$ is the strain hardening coefficient, ρ_0 is the initial dislocation density. This dependence is shown in Fig. 1.2.

Equation (1.4) and Fig. 1.2, unfortunately, define the function $\rho_m(\varepsilon)$ only qualitatively and are not suitable for real estimates, because the behaviour of dislocations during plastic deformation is much more complicated. This difficulty lies in the fact that, as shown by electron microscopic studies of thin foils cut from deformed crystals [Amelinks, 1968; Hirsch et al., 1968; Huvier et al., 2009; Hennecke, Hähner, 2009; Messerschmidt, 2010], in the course of plastic flow, bulk dislocation systems, dislocation ensembles, form in a crystal. Their combination forms an evolving dislocation substructure. An example of such dislocation ensembles, successively arising during the deformation of aluminium, is shown in Fig. 1.3, a–f.

The evolution of dislocation substructures with increasing strain can be represented, for example, in the following (not unique!) way [Kozlov, Starinchenko, Koneva, 1993]: the chaotic

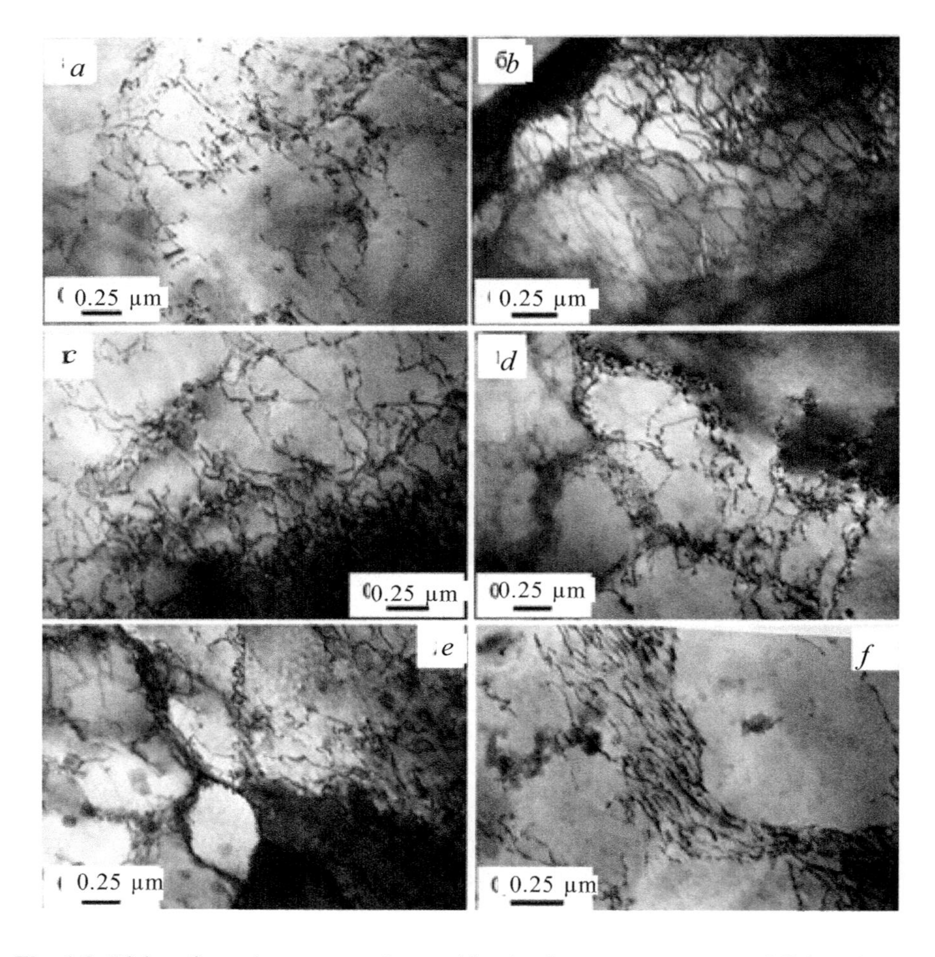

Fig. 1.3. Dislocation substructures observed in aluminum: *a* – structure of dislocation chaos; *b* – dislocation nets; *c* – bundles; *d* – cells; *e* – fragments; *f* – dislocation sub-boundary [Konovalov, 2012].

distribution of dislocations → dislocation clusters → homogeneous mesh substructure → dislocation coils → unoriented cells of the cellular-net unoriented substructure → cellular substructure with misorientation → cellular mesh substructure of smooth misorientation → band substructure → substructure with multidimensional disk and smooth misorientations → fragmented substructure.

The idea of complicating dislocation structures during deformation initiated a huge number of experimental studies of the dislocation structure [Rybin, 1986; Trefilov et al., 1987; Malygin, 1991a, b, c, 1995, 2004, 2005a, b, 2006; Mughrabi, 2004; Huvier et al., 2009;

Glezer, Metlov, 2010]. Their goal was to find the relationship between the parameters of the dislocation structure and deformation. A rather detailed picture of the evolution of dislocation ensembles during plastic flow was constructed, which turned out to be much more complicated than all the schemes incorporated in the original strain hardening models. For this reason, the main task of such studies has not been solved.

The large variety of dislocation substructures observed in different materials made their quantitative evolutionary description necessary to establish a connection with the shape of the plastic flow curve or, at least, with individual stages of strain hardening, virtually impossible. Actually, the theory of dislocations, when considered at the level of individual dislocations, turned out to be suitable at best for describing the phenomena of inelasticity and internal friction, when the deformations do not exceed 10^{-4}, and usually amount to (1-5) $\cdot$ 10^{-6}. Only for these cases it gives quite acceptable quantitative results [Granato, Lykke, 1969; Novick, Berry, 1975; Nikanorov, Kardashev, 1985].

1.1.2. Mesoscale strain level

This concept was introduced when it became clear that the development of a dislocation structure does not reduce to a monotonous increase in the density of dislocations, but is a complex process of forming new spatially inhomogeneous distributions of defects [Panin, Likhachev, Grinyayev, 1985]. For this reason, during the deformation, the material spontaneously stratifies into zones with different dislocation densities, in the distribution of which a clearly pronounced spatial periodicity ~$(10^2–10^3)$ $b \approx (10^{-8} – 10^{-7})$ m is observed. The limiting manifestation of this kind of plasticity heterogeneity is the occurrence of deformation-free channels that do not contain defects during deformation [Malygin, 1991*a, b*].

This led to the understanding of the fact that dislocation ensembles in the limit turn into independent defects that cannot be represented as a superposition of individual dislocations [Vladimirov, 1987]. Essentially, this is tantamount to recognizing the nonlinearity of a deformable medium [Aifantis, 2001; Kuhlmann-Wilsdorf, 2002]. The mesoscopic spatial scale was attributed to structural elements of this type [Panin, Likhachev, Grinyaev, 1985; Panin, 1998, 2000,

2014; Panin, Egorushkin, 2015]. This has led to the separation of such ensembles and the corresponding deformation phenomena in a separate area of research – the physical mesomechanics of materials. We note, however, that the above scale of phenomena in physical mesomechanics significantly exceeds the scale of ~ 10^{-9} m = 1 nm characteristic of mesoscopic physics [Imrie, 2002].

The first ideas about the existence of deformation defects larger than dislocations appeared long ago. So, Mott [Mott, 1951, 1952] at the early stage of the development of dislocation models introduced the concept of a flat cluster of dislocations, fundamental to the theory of dislocations, an idependent defect of dislocation origin that determines the nature of strain hardening of metals and alloys. Such a defect can occur when the source of Frank-Read dislocations is operating [Friedel, 1967]. A flat cluster has its own inhomogeneous elastic field [Stroh, 1954; Eshelby, 1963].

Mott's idea was successfully developed by Seeger [1960] and his students Berner and Kronmüller [1969], who based on it the theory of strain hardening of fcc single crystals at the stages of light slip and linear strain hardening. The advantage of this theory is that the strain hardening coefficient can be expressed in a dimensionless form through the ratio of two scales – the Burgers vector of dislocation and the length of a flat cluster. This conclusion from the Seeger theory is used now, almost seventy years after publication.

The tightening of plastic deformation conditions (high load rates, large deformations, typical, for example, of the rolling process [Srinivasa, 2001; Horstemeyer, 2002], complication of loading schemes and modes, low temperatures) led to the discovery of new types of plastic flow heterogeneity, such as shear bands [Gilman, 1994; Anand, Kalidindi, 1994]. These features have noticeably larger characteristic sizes, significantly exceeding the usual order for dislocations on the order of the Burgers vector. The plastic deformation microbands can have a non-crystallographic orientation, fundamentally differing from the slip bands in single crystals or individual polycrystal crystallites, where they are usually associated with densely packed planes with low indices. The spatial orientation of the microbands is determined by the deformation conditions: their slope to the normal to the direction of stretching is ~35° or ~50° under the condition of Tresco plasticity and ~35° under the condition of von Mises plasticity [Thomas, 1964; Rabbinov, 1988]. Anyway, in the case of microshift shears, we encounter a localized and non-uniform plastic flow.

For large strains, the description of the process is significantly complicated. So Glezer and Brooms [2010] showed that at a strain of ~1 (megastrain), plasticity carriers of various levels can serve as plastic carriers: point defects, dislocations, dislocation networks, grain boundaries of different types or their ensembles of a complex type. In this case, the structure of the carriers is continuously changing due to dynamic recrystallization.

Thus, the transition from the description of plastic flow using individual dislocations to a more realistic description of the description on a mesoscopic scale level turned out to be equivalent to the introduction of new specific defects of the crystal structure that are not reducible to individual dislocations, although obviously related to them.

1.1.3. Macroscale strain level

Turning to this scale, we note that in this case the characteristic dimensions of the inhomogeneity region and the sample are close. The most well-known and often discussed example of the localization of deformation on a macroscopic scale is the appearance and development of a neck of fracture at the final stage of the stretching process [Mac Lean, 1965; McClintock, Argon, 1970; Ekobori, 1971; Makhutov, 2005; Malygin, 2005a, b; Pelleg, 2013]. This well-known phenomenon is observed with the deformation and viscous destruction of ductile materials and manifests itself as a narrowing of the cross-section, which can reach tens of percent. The description of the material behaviour at this stage, when the conditional stresses in the sample decrease, signaling the loss of stability of the deformation process, is a serious scientific and technological problem [Struzhanov, Mironov, 1995].

Many macroscopic scale inhomogeneities have been known for a long time, and their nature has been well studied. These include the discharge bands, which usually arise during the compression of hcp and fcc crystals, accompanied by a lattice rotation [Honeycomb, 1972]. Macroscopic scale inhomogeneities also include twins, due to which many metals, alloys, and nonmetallic materials are deformed [Klassen-Neklyudova, 1960; Honeycomb, 1972; Kosevich, 1978].

Another type of macroscopic inhomogeneity of the plastic flow is known as plasticity waves [Kolsky, 1955; Shestopalov, 1958; Davis, 1961; Bell, 1984; Clifton, 1985; Meyers et al., 2001]. An experimental study of this phenomenon allowed us to establish that plasticity waves (stress waves in Kolsky's terminology [1955]) are macroscopic fronts of localized plastic flow moving along the samples under shock loading with velocity $V_{pw} \approx \sqrt{\theta/\rho}$, where ρ is the density of a substance. These waves can occur at strain rates of [Shestopalov, 1958]. Since $\theta \approx 10^{-3}$ G (G is the shear modulus), then $V_{pw} \approx \sqrt{\theta/\rho} \approx \sqrt{10^{-3}} \cdot \sqrt{G/\rho} \approx 0.03V_t \approx (30-100)$ m/s.

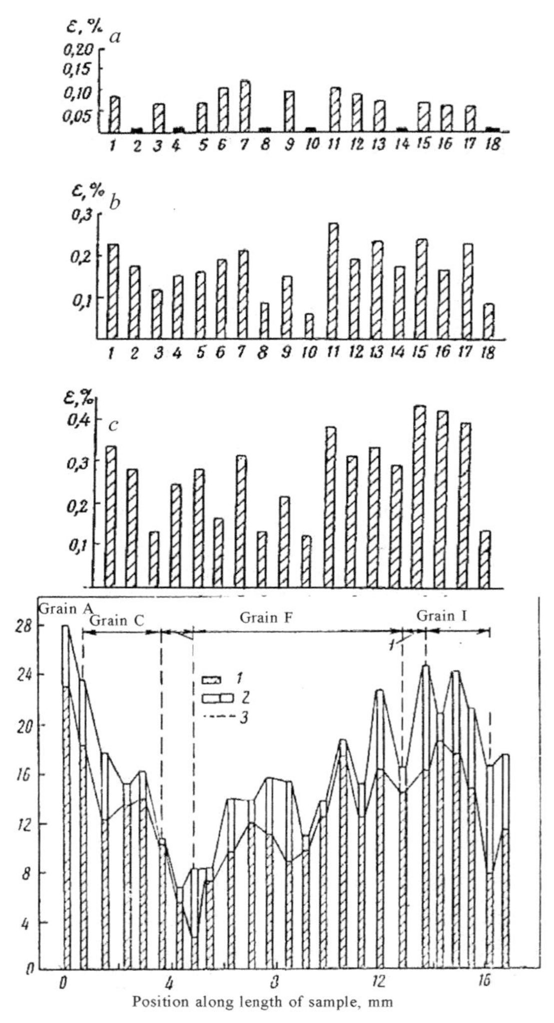

Fig. 1.4. Distribution of strain along the length of a steel sample for different times of deformation [Oding et al., 1959] (*a*); the same for polycrystalline aluminium [Garofalo, 1968] (*b*).

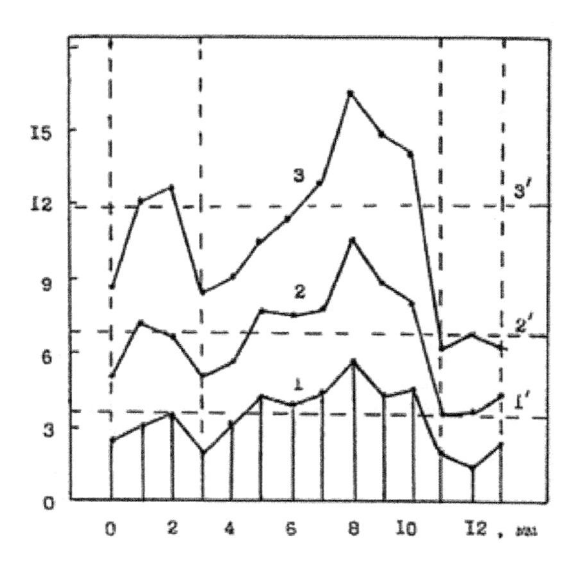

Fig. 1.5. Distribution of local strains along the length of the sample in Al [Kibardin, 2006]. Total deformation of sample 0.037 (1), 0.069 (2), 0.11 (3).

A well-known example of the macroscopic localization of plastic flow are Chernov–Lüders bands [Mac Lean, 1965; Honeycomb, 1972; Pelleg, 2013], the observation of which marked the beginning of a modern approach to the study of plasticity [Lüders, 1860; Chernov, 1950]. The Chernov–Lüders band is a zone of plastic deformation originating in an elastically intense sample, expanding due to the movement of its borders (fronts). In this mode of deformation, which is usually the yield point on the $\sigma(\varepsilon)$ diagram, the fronts separate the areas of elastic and plastic deformation from each other. In the sample, several bands can develop simultaneously. The basis of the dislocation models of the band, described in [Christ, Picklesimer, 1974; Fujita, Miyazaki, 1978; Andronov and Gvozdikov, 1987; Sun et al., 2003; Plekhov et al., 2009; Petrov and Borodin, 2015], lies the consistent spread of dislocation shifts from one grain to another. Gehner [Hähner, 1994] combined two approaches to the problem of the Chernov–Lüders band and first introduced the concept of solitary plastic waves for these purposes.

If the patterns listed above relate *in situ* to the observed features of macroscopically localized plastic deformation, then there are indirect indications of macrolocalization, noted, for example, by Oding et al. [1959] and Garofalo [1968] (Fig. 1.4a, b). The localization of plastic flow in these cases was revealed by measuring the distribution of strains after the completion of the experiment. The results of these studies, presented in the form of diagrams of the distribution of plastic deformation along the length of the sample, an example of which is shown in Fig. 1.4a, b, demonstrate the features of this phenomenon. Attention is drawn to a certain periodicity in the distribution of local deformations, which increases during the course of deformation.

Detailed studies of plastic flow heterogeneity at the macroscopic scale level in technical-grade aluminium and alloys based on it were conducted by Weinstein, Kibardin and Borovikov [1982], Kibardin [1981, 2006], and Weinstein and Borovikov [1982]. In these experiments, the spatial periodicity of the distributions of local deformations turned out to be quite noticeable. It is so clearly manifested in the cyclical change in the autocorrelation two-dimensional function of aluminum microstrains, as shown in Fig. 1.5 that Kibardin [1981] found it possible to treat the observed changes as a deformation wave.

In the course of statistical data processing, the parameters of plastic flow inhomogeneity were determined and strain distributions were studied in individual elements of the sample [Weinstein, Kibardin and Borovikov 1982]. These techniques gave an average estimate of the degree of localization at each degree of deformation of the material, before the appearance of neck fracture. As a result, there was a strong smoothing of temporary changes in microstrains and, in essence, a transition from the strain field to an integral parameter generalized for the whole object, which characterizes the degree of strain localization.

Cinematographic studies of the change in the shape of the loaded specimen carried out by Wray [Wray, 1969, 1970] on stainless steel and Pb–In alloy showed that, starting with 1% deformation, several necks of cylindrical specimens of Pb–In alloy arise from uniaxial stretching. They can be active (the strain rate in them is higher than in the whole specimen) or inactive. First, in the process of stretching, the number of necks, including active ones, increases. In this case, the strain rate in them can vary. At total strains of ~0.3 ... 0.4, one of the necks becomes stable, causing further destruction. The observed changes in the activity of the necks prove that in the experiments it was possible to fix the cyclic spatio-temporal process of evolution of the distributions of local deformations. As in the works of Weinstein, Borovikov [1982] and Kibardin [2006], the conclusion about the wave nature of the deformation in the works of Ray was not made because the instability of the plastic flow was considered a priori as a random phenomenon, not subject to any spatio-temporal laws.

In studies of the localization and instability of plastic flow performed by Presnyakov [1981] and Presnyakov and Mofa [1981], photographing the sample during deformation was also used. These authors noted that, in the process of deformation, long before the emergence of a stable neck, zones of deformation localization moving along the sample are formed, which they call running necks. The authors managed to record the oscillations of local and maximum strain rates under tension at a constant speed of aluminium and its alloys. The spatial heterogeneity of the plastic flow, the displacement of the zones of localization of deformation and the oscillation of the velocities of the plastic flow also indicated the occurrence of the wave process during plastic deformation. Unfortunately, the authors did not provide data on the speeds of movement of the moving necks and, in general, no quantitative data on the spatiotemporal distributions of local macrostrains.

It is also worth mentioning the work of Golubev [1950], in which it is shown that the development of plastic deformation during rolling in rolls of large diameter looks like a gradual spread of the front of plastic deformation along the section of the ingot from the surface inward.

Thus, localization of the plastic flow at the macroscopic level is also accompanied by the formation of features in the distribution of deformation, which, apparently, cannot be represented as the sum of simpler defects. It can be assumed that in all these cases we encounter manifestations of the nonlinearity of a plastically deformed medium [Aifantis, 1996; Scott, 2007], who emphasize the complexity of the behaviour of a continuous medium during its plastic deformation.

1.1.4. Lattice scale level

Usually, when discussing the nature of plastic deformation, elastic processes in the crystal lattice are practically ignored. This is not quite fair, if only for the reason that elastic, inelastic and plastic deformation are closely interconnected, change continuously into each other, as indicated by the continuous view of the flow curve $\sigma(\varepsilon)$ [Hill, 1956; Novick, Berry, 1975; Landau, Lifshits, 1987]. At the lattice level, elastic strains associated with small distortions of the ideal crystal lattice are usually considered. However, areas of balancing internal stresses should also be considered as areas of localization of elastic deformation [Barrett, 1948].

The most interesting is the possibility of localization of high-amplitude elastic waves. This question was considered in detail by Porubov [2009], who showed that at short-term effects of sufficiently high amplitude in a deformable medium the localization of elastic waves can occur, taking on different forms (see also [Friedlander, 1962]).

Finally, the nonlinearity of the elastic medium leads to the possibility of the appearance of soliton-type perturbations in it under external influence [Rabby, 1980; Dodd et al., 1988; Kosevich, Kovalev, 1989; Kerner, Osipov, 1989; Zakharov, Kuznetsov, 2012; Maugin, 2013], which are more complex than localized elastic waves formations. We also take into account that the Frenkel–Kontorova dislocation [Kontorova, Frenkel, 1938], whose introduction marked the beginning of the theory of dislocations, can also be considered as a soliton [Brown, Kivshar, 2008].

Apparently, all variants of plastic flow localization are associated with the inherent property of the crystal lattice to exhibit instability during deformation, accompanied by the destruction of the body or the birth of crystal defects [Thompson, 1985; Krempl, 2001; Grimwall et al., 2012].

1.1.5. Temporal nonuniformity of plastic flow

In addition to spatial heterogeneity (localization) of the plastic flow, the latter is also characterized by temporal heterogeneity, usually manifested as a jump-like deformation or the Portevin–Le Chatelier effect [Bell, 1984; Zaiser, Hähner, 1997; Khristal, 2001; Rizzi, Hähner, 2004; Pelleg, 2013]. In this case, the plastic flow curve has a sawtooth-like appearance, and the load breakdowns occurring over short periods of time are clearly seen on it. The discontinuity is characteristic of a number of pure metals and alloys, as well as non-metallic crystals, especially filament-like ones [Zuev, 1990]. At the same time, a decrease in the test temperature and an increase in the rigidity and sensitivity of the testing machine contribute to the manifestation of jump-like deformation. For example, at the temperature of liquid helium (~4 K) pure aluminium, for which the plastic flow curve $\sigma(\varepsilon)$ is smooth under normal conditions, demonstrates pronounced jumps in the deforming stress (Fig. 1.6, b; [Pustovalov, 2008]).

WWhe nature of the effect is usually associated [Bell, 1984; Khristal, 2001; Rizzi, Hähner, 2004] with strain-induced ageing caused by the formation of impurity atmospheres [Cottrell, 1958] on dislocations, which were detained during their movement by local barriers of different nature [Wirtman, 1987; Zbib, de la Rubia, 2002]. To continue the deformation, it is necessary to increase the effective stresses in order to ensure the separation of dislocations from the pinning points.

An alternative point of view on the nature of the discontinuous deformation, developed by a number of authors [Nagorny, Sarafanov, 1993; Shibkov, Zolotov, 2009; Shibkov, Zolotov, Yellow, 2012; Shibkov et al., 2016; Shibkov et al., 2014; Shibkov et al., 2016], is based on the instability of the plastic flow caused by the extreme dependence of the density of mobile dislocations on the strain [Gilman, 1965 and 1972]. In this case, with the release of dislocations from the pinning points, the density of mobile dislocations increases abruptly. In accordance with equation (1.1), to continue the deformation at the same speed, the velocity of motion of dislocations should fall, which is possible only with a decrease in the deforming stress. There are many works where the effect is explained from this point of view [Kubin, Estrin, 1985, 1992; Kubin, Chinab, Estrin, 1988; Zaiser, Hähner, 1997; Rizzi, Hähner, 2004; Zuev, 2017a, b]. Such an approach makes it possible to achieve a more accurate quantitative description of the phenomenon of abrupt deformation.

At low-temperature deformation, when diffusion phenomena are suppressed and thermally activated detachments of dislocations from pinning points are unlikely, the effect is usually explained by local heat generation on slip planes due to the adiabaticity of shear processes. Due to the low thermal conductivity of metals at such temperatures, thermal energy does not have time to dissipate during the course of shear. Local heating of the material in the slip plane increases the likelihood of thermally activated acts in this zone to overcome obstacles while reducing the deforming stresses necessary for this purpose [Klyavin, 1987; Dotsenko, Landau, Pustovalov, 1987; Nicolis, Prigogine, 1990; Rizzi, Hähner, 2004; Pustovalov, 2008].

1.2. Models of different-scale processes of plastic flow

One of the most significant achievements in the physical theory of plasticity in recent decades has been the introduction and development of the concepts of plastic flow as a multilevel process [Panin, Likhachev, Grinyaev, 1985; Likhachev, Malinin, 1993; Zbib, de la Rubia, 2002]. In this approach, the spatial localization and temporal heterogeneity of the plastic flow associated with the birth and evolution of lattice defect ensembles can be viewed as different collective modes of the form-changing process [Aifantis, 1992, 2001; Naimark, 1997, 2003; Borg, 2007]. Recognition of the existence of such collective formations and the fundamental difference in their scales have posed to the researchers the problem of quantitative and qualitative coordination of such different scales of simultaneously or sequentially occurring phenomena and the hierarchy of their organization.

1.2.1. Dislocation models

Already starting with the first works in the field of strain hardening theory, working models and hypotheses take into account the heterogeneity of the distribution of defects in a deformable medium and the tendency of plastic deformation to localization. The main role in most models is played by geometrically relatively simple dislocation assemblies such as dipoles, flat clusters, and dislocation walls (low-angle boundaries) [Seeger, 1960; Mughrabi, 1983, 2001, 2004; Mughrabi, Pschenitzka, 2008; Nabarro, Bazinsky, Holt, 1967; Nabarro, 2000; Landau et al., 2009; Messerschmidt, 2010]. Within the framework of their interaction, it is possible to describe various elementary acts of plastic flow and fracture of crystals.

The development of plastic deformation of solids is determined by both externally applied forces and the interaction of defects of the crystal structure generated in this process. Due to this, collective effects arise in the system of defects, gradually leading to the formation of plasticity carriers that are structurally more complex than individual dislocations [Vladimirov, 1987; Wirtman, 1987; Kuhlmann-Wilsdorf, 2002]. In fact, this process proceeds from the very beginning of the deformation, since by the end of the easy slip stage in crystals there are flat clusters of dislocations and more complex ensembles of defects. For this reason, approaches that have proven themselves well in describing the initial stages of plastic flow turn out to be practically useless when analyzing developed plastic deformation.

As a result of the action of these two reasons, the real elementary acts of the process of change become no longer associated with individual dislocations, but with their ensembles — dislocation formations that serve as carriers of the deformation under the conditions considered. Obviously, in this case, the characteristic size (scale) of the volume, within which a single plasticity event occurs, increases compared with the case when the deformation can be described in the approximation of individual dislocations without taking into account collective effects in the dislocation subsystem of the crystal [Eshelby, 1963; Kuhlmann-Wilsdorf, 2002].

In general, it can be assumed that each deformation mechanism corresponds to a similar characteristic size, which depends, firstly, on the type of plastic deformation carrier (point defects, dislocations, twins, defect ensembles, etc.) [Orlov, 1983; Vladimirov, 1987], secondly, on their mobility and free path [Dotsenko, Landau, Pustovalov, 1987] and third, on the presence of collective effects and, accordingly, on the mobility and path length of elementary carriers formed due to this [Aifantis, 1984, 1995; Borg, 2007; Landau et al., 2009]. Thus, at high degrees of general deformation or when it is strongly localized, a situation arises when analyzing several types of acts that are 'elementary' for the chosen working volume can be distinguished. Accordingly, as shown by Vladimirov and Kusov [1975], as well as Vladimirov [1987], this requires the introduction of several effective scales for the description of the form-changing process that looks uniform.

Such a conclusion turned out to be fundamentally important and made it possible to formulate the principle of the multilevel nature of plastic deformation [Panin, Egorushkin, 2015]. Its consistent use leads to the conclusion that, although the phenomena at all physically real levels are interdependent and interrelated, a selective approach to the study of processes characteristic of each of them is possible.

1.2.2. Large-scale distribution of strain

The most consistent attempt to concretize ideas about the multilevel nature of the process of plastic flow was undertaken by Likhachev and Malinin [1993]. Seeing the main task of the concept of structural levels in ensuring the possibility of engineering calculations of deformation at the macroscopic structural level, they proposed an algorithm for choosing the elementary level for each specific case of plastic deformation. For this, in their opinion, it is necessary:

- that the considered act of plasticity is independent of similar occurring in other parts of the object;
- so that it admits an invariant notation of relations that define this process;
- so that the lowest level could be as large as possible to overcome computational difficulties when moving to the engineering level.

Likhachev and Malinin [1993] believed that in such a formulation the lower structural level cannot be really elementary (atomic or, at least, dislocation), but its systematic description can be rigorously justified. As an example, the authors considered the level of slip bands for materials where they are observed. Indeed, a thermal-fluctuation or force deformation mechanism with a linear law of strain hardening acts in the slip band. Under these conditions, they were able to strictly write down the governing equations for total strains and stresses. The transition to the next structural level should be carried out by orientation and statistical averaging.

However, such an averaging procedure, as shown by Selitser [1989], is nontrivial, since it is necessary to take into account the strong interaction between lower-level defects. On the basis of the chosen interaction laws, one can obtain structural-orientational functions that provide the transition to the second, higher level of deformation, which can already be considered as macroscopic. This concept, despite its allure, is confronted with a number of difficulties. For example, it is not always clear how, in the general case, it is possible to specify the interaction of defects at the lower level. Further, in the process of deformation, a change in the mechanism of the elementary plasticity event may occur, which will lead to the need to replace the selected lower level, and, consequently, the entire calculation procedure. This also leads to a change in the temperature conditions of deformation. However, even with these difficulties in mind, this concept of structural levels has proven to be productive for solving a wide range of tasks.

Vladimirov and Kusov [1975], Vladimirov and Romanov [1986], and Rybin [1986] associated plastic flow transitions from one level to the next, more large-scale, with the initiation of rotational modes of plastic deformation. In their opinion, the transition to collective effects and the inclusion of relevant processes at the higher scale level is caused by the achievement of the critical density of deformation defects acting at the lower scale level.

The same ideas were used by Zasimchuk [1989], showing that the instability of the properties of deformed crystals is determined by the fact that a deformable crystal is an open thermodynamic system, the flow of energy through which causes its self-organization. This view is consistent with the general concepts of the nature of self-organization in open systems [Krinsky, Zhabotinsky, 1981; Nicolis, Prigogine, 1979; Othmer, 1991; Pontes, Walgraef, Aifantis, 2006; Haken, 2014].

Meyers et al. [Meyers et al., 2001] used a multiscale approach to analyze the phenomena that determine the evolution of shear bands under explosive loading. In simulating three-dimensional plastic flow problems, they were able to propose principles for describing multiscale deformation near the sample surface, as well as for the formation of microband shear bands and dislocation boundaries.

Thus, there is reason to believe that the enlargement of the current scale of information phenomena takes place in the course of a multistage process of plastic flow. In this case, the possibility of simultaneous implementation of multi-scale deformation processes is not excluded.

1.3. Plastic deformation as self-organization

In 1987, Seeger and Frank [Seeger, Frank, 1987] published an article

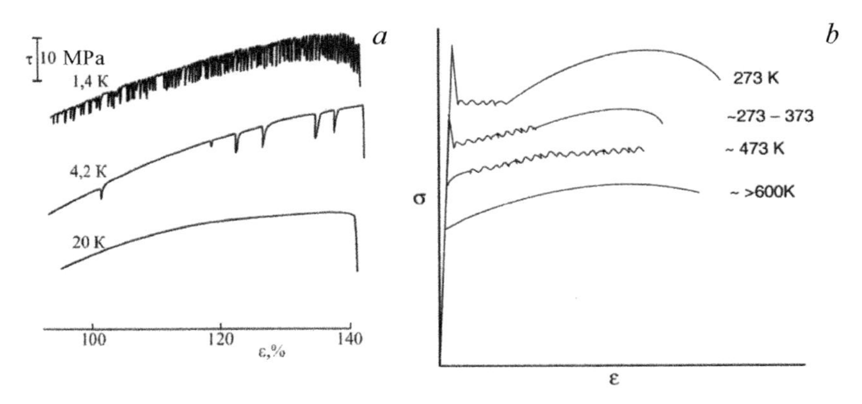

Fig. 1.6. Examples of plastic plastic deformation: deformation scheme during the Portevin effect–Le Chatelier effect [Pelleg, 2013] (a); hopping deformation of pure Al [Pustovalov, 2008] (b),

in which they attributed the processes occurring in the medium during deformation to structure formation. The list of literature used by the authors included the famous names of Ebeling, Hacken, Nicolis and Prigogine, whose works were not previously quoted in works on the strength and plasticity of solids.

The listed authors are the creators of synergetics – the science of education structures [Ebeling, 1979; Hacken, 1985, 2014; Nicolis, Prigogine, 1979, 1990], so the work of Seeger and Frank attracted the attention of specialists in the field of plasticity to synergetics. Interest in this area was inspired, in all likelihood, by the fact that, generally speaking, synergetics was born from the controversy of two fundamental laws of nature [Glensdorf, Prigogine 1973; Ebeling, 1979]:

- the second law of thermodynamics, which requires an increase in entropy and destruction of the structure existing in the system (Clausius),
- evolutionary theory, which provides for the complexity of the structure of the system as it develops with a local decrease in entropy (Darwin).

In plasticity physics, a similar situation developed: thermodynamics required the degradation of the properties of a deformable medium, and the experiment unambiguously pointed out the complication of its structure during deformation.

1.3.1. On the possibilities of synergetics in the theory of plasticity

By the 80s of the 20th century, the main principles of synergetics were formulated on the basis of the thesis of Glansdorf and Prigogine [1973], according to which: "*Classical thermodynamics, in essence, is a theory of structural destruction. But classical thermodynamics must somehow be supplemented by the structure creation theory missing in it*".

Nicolis and Prigogine [1990] were practically the first to realize that "*Mechanical phenomena ... should be considered as part of the general problem of nonlinear dynamic systems operating far from equilibrium.*" The ideas about structure formation with plastic flow suggested by Seeger and Frank [Seeger, Frank, 1987] gave the opportunity to use the ideology and apparatus of synergetics, which by this time reached a sufficiently high level (see, for example, the books of Glansdorf and Prigogine [1973], Nicolis and Prigogine [1979, 1990], Haken [1985, 2014], Chernavsky [2004]). In addition, since these books deal with issues that are far from the problems of the plasticity of solids, there is a useful opportunity to borrow analogies from chemistry, biology, and even enomics (see, for example, [Chernavsky, Starkov, Shcherbakov, 2002]), which is based on the universality of the mathematical apparatus of synergetics.

If until this time in the physics of plasticity, the rule "*structure determines properties*" was used [Honeycomb, 1972; Pelleg, 2013], it has now become clear that, on the contrary, "*the process of plastic flow determines the structure of a deformable medium.*" For this reason, collective modes of plastic deformation of steel associated with structure formation have been considered within the framework of synergistic patterns as done by Aifantis [Aifantis, 1994, 1995, 1996], Landau et al. [Landau et al., 2009], Maugin [Maugin, 2013].

It became clear that there are internal and external causes of the appearance of collective effects in the dislocation structure of

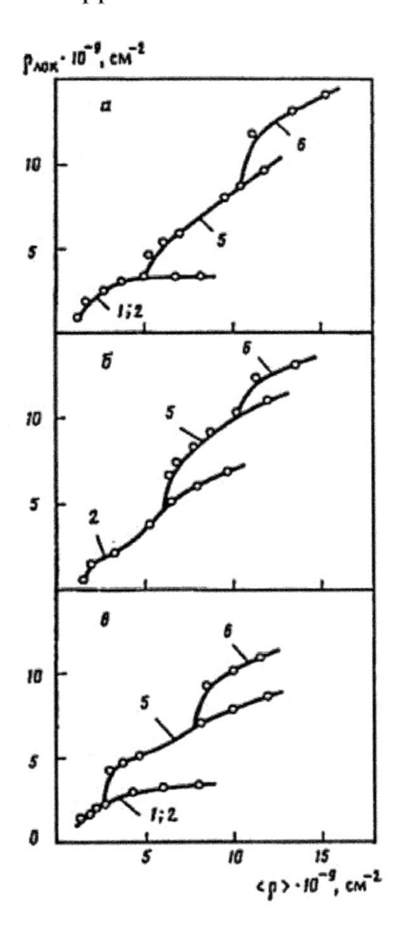

Fig. 1.7. Dependence of the local dislocation density on the average scalar density in copper alloys. Cu–5 at.% Al (*a*); Cu–0.5 at.% Al (*b*); Cu–0.4 at.% Mn (*c*). The figures for curves are the types of substructures: 1– random, 2 – clusters, 3 – cellular, unoriented, 6 – cellular, disoriented [Kozlov, Starenchenko, Koneva, 1993].

a deformable crystal. The internal reason is connected with the geometry in the mutual arrangement of dislocations. At a dislocation density of $\sim 10^{12}$ cm^{-2}, the effective stress of their interaction is $\sigma_{id} \approx Gb\sqrt{\rho} \approx 10^{-2}G$ [Friedel, 1967; Wirtman, 1987] is an order of magnitude larger than the external deforming stress of $\sim 10^{-3}G$. Under such conditions, the decisive role can be played by the relaxation processes of the formation of ordered dislocation substructures with a low elastic energy [Krempl, 2001; Kuhlmann-Wilsdorf, 2002; Veyssiére, 2009].

The external cause of collective effects is of thermodynamic origin and is determined by the fact that the deformable sample is not an isolated system: during deformation, energy flows from the loading device through it. A similar situation is characteristic of the thermodynamics of nonequilibrium processes [De Groote, Mazur, 1964] or in a more general formulation for the problems of synergetics [Hacken, 1985, 2014; Nicolis, Prigogine, 1979, 1990] and is considered as a necessary condition for the spontaneous self-organization of the environment [Mishchenko et al., 2010].

1.3.2. On collective phenomena in plasticity

A new approach to the explanation of the laws of plastic flow, considered in the reviews by Malygin [1995, 1999], began to take shape as these circumstances were understood. Within its framework, deformation is considered as a collective process, during which different types and complexities of plasticity are generated, regularly evolve and destroyed. In a number of works ([Malygin, 1991, 2901, 2004, 2005, 2011; Hannanov, 1992; Hannanov, Nikanorov, 2007; Kubin, Estrin, 1985, 1992; Sarafanov, 2001, 2008; Nagorny, Sarafanov, 1993; Maksimov, Sarafanov, Nagorny, 1995; Zaiser, Hähner, 1997]) it was possible to solve residual complex problems and obtain a description of the processes of solidification of solids under loading.

The evolution of dislocation substructures during deformation was investigated by Kozlov, Starinchenko, Koneva [1993], who experimentally established that the restructuring of one type of defect structure in another has signs of a phase transition. In the course of such a restructuring process, a new dislocation substructure arises and begins to develop against the background of a simpler, already existing, but completing its evolution, as shown in Fig. 1.7.

The next logically necessary step in the development of a synergistic approach to the nucleation and evolution of the inhomogeneous distribution of defects in the course of plastic flow was to take into account the nonlinearity of the deformable medium. Attracting this concept is quite natural with respect to a deformable medium, since the crystal 'remembers' everything that occurs in it at stresses above the yield strength due to irreversible changes in its defective structure.

It is known that the nonlinearity of a plastically deformed solid medium manifests itself immediately upon reaching the yield strength [Aifantis, 1996; Olemskoi, Sklyar, 1992; Olemskoi, Khomenko, 2001] and is usually expressed in the fact that the properties of a deformable material are largely determined by its prehistory and the nature of the defective subsystem arising at the previous stages of the deformation process [Alshits et al., 2017].

Accounting for nonlinear effects in plasticity physics is complicated, and only in a small number of cases was this problem successfully solved. For example, Nazarov [2016] modernized the theory of internal friction proposed by Granato–Lücke [1969] and predicted the existence of a number of nonlinear effects of the interaction of elastic waves in a medium containing defects.

Kiselev [1999, 2006] considered the occurrence of travelling cnoidal waves (cnoidal waves are solutions of the Korteweg – de Vries equation [Dodd et al., 1988]) taking into account the nonlinearity of dislocation ensembles and suggested a mathematical interpretation of their origin and evolution. He showed that the cause of the periodic nature of plastic deformation is the interaction between defects, increasing with increasing strain.

The most successful attempts at the macroscopic description of nonlinear plasticity should, of course, include the gradient theory of plasticity developed in the works of Aifantis [Aifantis, 1999, 2001, 1996, 1984, 1992, 1995, 1994, 1987], Zaiser and Aifantis [Zaiser, Aifantis, 2006], Pontes, Walgraef and Aifantis [Pontes, Walgraef, Aifantis, 2006], as well as his followers [Abu Al-Rub, Voyiadjis, 2006], [Borg, 2007]. This theory, taking into account the multi-scale deformable medium by introducing the internal scales of different sizes, allows to achieve a satisfactory quantitative agreement with the results of experimental studies of various stages of strain hardening of materials.

Another approach to the problem of inhomogeneous plastic deformation has been developed in the works of Naimark [Naimark,

1997, 1998, 2003, 2015; Naimark, 1997, 2003], Naimark, Ladygina [1993] and Naimark, Davydova [Naimark, Davydova, 1996]. The authors showed that the nucleation and evolution of defects during plastic flow causes a change in the local and global symmetry of the system, similar to the transition from laminar to turbulent flow in hydrodynamics [Klimontovich, 2002]. In this case, a local change in symmetry is associated with the inhomogeneity of the evolving field of the components of the plastic distortion tensor, and the global one is the change in the symmetry of the entire system when collective plastic deformation modes are generated and the behaviour of the system is subordinate to these modes. In the framework of this approach, the localization of plastic deformation is considered as an orientational kinetic transition, accompanied by spatial-temporal self-organization of defects. In this case, the cause of the instability of the process and the occurrence of plastic flow localization is the interaction of individual defects (dislocations), leading to the emergence and further evolution of more complex ensembles of defects.

1.4. Plasticity problem

If, on the basis of the analysis of existing experimental and theoretical data on the physics and mechanics of plastic deformation of solids, we try to identify the most significant regularities in the development of this phenomenon, it is easy to see that there are two regularities:

- plastic flow is localized at all levels of its development,
- plastic flow develops by complicating the defect structure of the deformable medium.

We consider these patterns sequentially.

1.4.1. Localization and the self-organization of plastic flow

A general acquaintance with the results of studies of the processes of plastic deformation of crystalline solids shows that the tendency to localization remains one of the most important, but still not completely understood problems, inextricably linked to plasticity.

It is fundamental that strain localization is observed at all scale levels of this process, as was shown above. Even at the lattice scale level [Porubov, 2009], the propagation of elastic waves of sufficiently large amplitude is also accompanied by their localization and the formation of wave features due to the nonlinearity of the elastic medium [Ashcroft, Mermin, 1979]. As for dislocations, already in the first works devoted to their properties [Kontorova, Frenkel, 1938; Reed, 1957] (see also [Brown, Kivshar, 2008]), it was shown that the shift associated with breaking and restoring interatomic bonds is localized on the acting slip planes, while the material between these planes remains elastically deformed. In this case, the shift caused by one dislocation is equivalent to the displacement of one part of the crystal relative to another by the magnitude of the Burgers vector b of the dislocation. It follows from this that plastic deformation at the dislocation scale level is localized.

The same reasoning can also be attributed to deformations encompassing meso-and macroscopic scale levels. Essentially, all the features of the distribution of deformation defects in these cases indicate areas where plastic deformation occurs (localized) at the time of observation.

In the general formulation, the localization problem has both scientific and practical aspects. The scientific side of the problem attracts with its generality and mystery. Localization occurs spontaneously, and in a deformable material in the initial state there may be no heterogeneity of composition or structure, with which it would be natural to associate the occurrence of localization. The practical importance of this problem is also difficult to overestimate, since localization prevents the plastic flow from being necessary for many technological processes and makes the required high degree of plastic deformation to failure unattainable [Frost, Ashby, 1989; Breeders, Zuev, Kotrekhov, 2012].

The complication of the deformation structure during the deformation process is clearly subject to certain laws, which are currently trying to explain in the framework of the theory of non-equilibrium structures (synergetics) [Zaiser, Hähner, 1997; Olemskoi, Khomenko, 2001; Aifantis, 2001; Olemskoi, Katznelson, 2003; Zaiser, Aifantis, 2006]. These patterns control transitions from lower to higher levels during plastic deformation. It must be admitted that to date attempts to use synergistic ideas in the physics of plasticity have not yet led to the creation of any reliable theory of structure formation. The reason for this is, apparently, the absence of a systematic approach to the problem of describing plasticity in terms of synergetics. In addition, the principles of choosing the laws of plastic flow processes that could become objects of synergy of plasticity are still unclear.

1.4.2. On the principles of constructing a model of plastic flow

Analysis of the current situation in the field of studies of plastic deformation patterns shows that there is a fairly large amount of experimental information on localization at the microscopic (dislocation) scale level, and there are also physically based models describing the occurrence of dislocation substructures that are more complex than with individual dislocations (mesoscopic scale level). At the same time, the phenomenon of localization of the plastic flow of a macroscopic scale remains largely mysterious, both in the phenomenological sense and in the sense of understanding the nature of the mechanisms underlying it. Both of these problems obviously require special attention.

For these reasons, it appears that the construction of a new model of plastic flow should be based on the following provisions:

- Experimental evidence of the localization of plastic flow as an indispensable attribute of plasticity, having different forms and scales of realization, should be put at the forefront,
- within the framework of the developed model, the localization of deformation should be considered the result of self-organization (reduction of symmetry) of the deformable system in accordance with the laws of synergetics,
- to introduce the time factor into the plastic flow model, it follows by analogy with the theory of elasticity, in which time is introduced via the elastic wave equation, to identify and use any wave process related to plasticity,
- the mathematical form of the model is determined by the fact that *for the occurrence of the phenomenon of spontaneous symmetry breaking with a decrease in its degree, the system must be open, and its mathematical model must be nonlinear* (Mishchenko et al. [2010]).

There is also the paradoxical view that the required model should be rather 'rough'. In this case, it retains its resistance to random variations of variables and can correctly describe the laws of the process being modeled. As Chernavsky [2004] stated, *if a model is rather coarse, then small factors do not distort its results; if the model is non-rubbish, then it cannot describe reality.* Arnold [1998] formulated this idea in the following way: *Complex models are rarely useful (unless for the dissertation studies).*

These are the reasons that prompted the author to attempt to investigate and describe the process of plastic flow within the framework of a tendency opposite to the generally accepted one, that is, moving in the direction from the macroscopic properties of a deformable medium to its microscopic properties. The results of these studies are the subject of this monograph.

2

Macroscopic localization of plastic flows

The literature data discussed in Chapter 1 convince us that the main regularity of the plastic flow process is its tendency to a non-uniform distribution of deformation over the volume of a deformable body, that is, to localization of plastic deformation, which manifests itself at all stages of the process. For this reason, it seems that sufficiently complete information about the laws of localization processes can serve as the key to solving the problem of plastic flow. It is also clear that the success of an experimental study of the problem of localization is determined first of all by the availability of a sensitive technique, which makes it possible to observe the spatial–temporal characteristics of the deformation field in real time (*in situ*).

2.1. Methods of observing patterns of localized plasticity

The requirements for the method of studying the localization of plastic flow in deformable materials are key when searching for the main laws of plastic flow. These requirements are almost obvious: the technique should combine the accuracy of measuring the quantitative characteristics of the plastic flow process at the resolution level of an optical microscope ($\sim 10^{-6}$ m) with a field of view of the order of the sample length ($\sim 10^{-1}$ m). The first requirement is necessary to ensure sufficient sensitivity of the device, and the second is dictated by the need to observe the deformation of the sample as a whole.

Creating such a technique is a nontrivial task, but it is known that a suitable combination of characteristics is achieved in the holographic interferometry method [West, 1982; Schumann, Dyuba, 1983; Ostrovsky, Shchepinov, Yakovlev, 1988; Kudrin, Bakhtin, 1988]. In our works, a variant of the holographic technique was used, based on the use of the speckle effect (the laser spot effect) [Franson, 1980] and adapted to solve the problems of deformation mechanics. The speckle effect is that when the scattering surface of an object is illuminated with coherent laser light, its photographic image becomes grainy because the coherent light scattered by a point on the surface interferes with the light scattered by other points. As a result, a chaotic pattern in space is formed, but because of the high temporal light of the light, the interference pattern that is stationary in time is a system of bright spots – speckles separated by dark gaps. The distribution of speckles in the image is determined by variations of the microrelief, varying from point to point. West [1982], Klimenko [1985], Jones and Wykes [1986], Kudrin, Bakhtin [1988], and Rastogi [1997] developed methods for applying speckle effect in metrology, based on a comparison of two speckle images, one of which was obtained before displacement or deformation of the object, and the second – after. The comparison can be realized by photographic or electronic means.

To study the plastic deformation kinetics, the ALMEC and ALMEC-tv (Automatic Laser Measuring Complex) research complexes were created at the Institute of Strength Physics and Materials Science of the Siberian Branch of the Russian Academy of Sciences, which are based on the speckle effect. They were used to analyze the laws of plastic flow in metals and alloys [Zuev, Danilov, Barannikova, 2008], alkali-halide crystals [Barannikova, Nadezhkin, Zuev, 2010; 2011], ceramics [Barannikova et al., 2007] and rocks [Zuev et al., 2014], as well as for solving scientific and technological problems (see, for example, [Zuev et al., 2002; Breeders, Zuev, Kotrekhov, 2012].

The main functional characteristics of the complexes are given in Table 2.1. Further, their designs and principles of operation are considered separately for each of the devices.

2.1.1. ALMEC complex. Principle of operation

In this complex, photographic registration of speckle images of flat samples and the optical method of their interpretation are used. The work of the complex in studying problems of plasticity is described in detail in the book by Zuev, Danilov, Barannikova, [2008], as well as in the articles by Gorbatenko, Polyakov, Zuev [2002] and Polyakov, Bikbaev, Zuev [2004]. The scheme of photographic registration of images is shown in Fig. 2.1 a. A He–Ne laser with a wavelength 32 λ_{las} = 632.8 nm is typically used to illuminate the scene under study. When performing measurements on the same photocarrier, two images of an object superimposed on each other are sequentially recorded, with the first exposure being made before deformation by a specified value, and the second after it. The resulting image of the sample is modulated by two speckle systems, each of which corresponds to one of the surface points and serves as its optical mark. The speckle patterns superimposed on the image of an object are shifted relative to each other by the displacement vector r, which is generally a function of the coordinates $r = r\ (x,\ y,\ z,)$ Double speckle pictures are called two-exposure speckle photographs or specklograms [Klimenko, 1985].

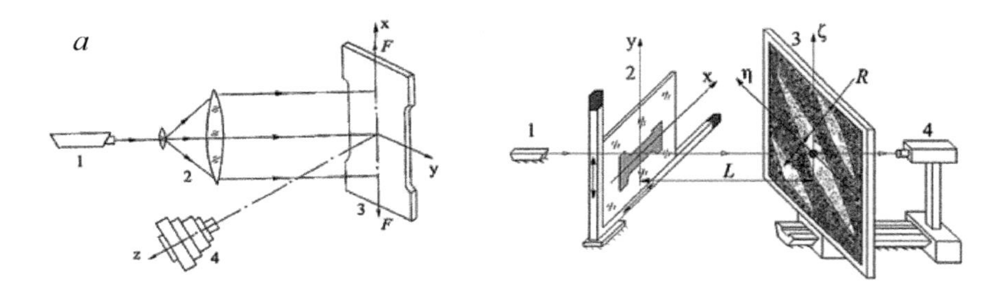

Fig. 2.1. ALMEC complex. The registration pattern of specklograms (a): 1 – laser; 2 – collimator; 3 – object; 4 – camera; specklogram (*b*) decryption scheme: 1 – laser; 2 – device for positioning and moving speckle photographs; 3 – screen; 4 – video camera,

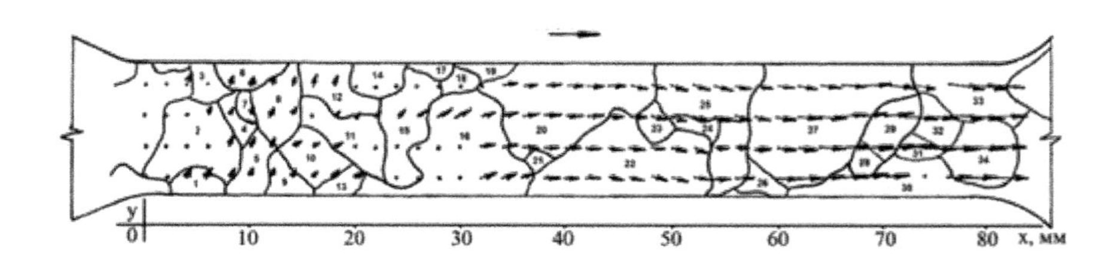

Fig. 2.2. The field of displacement vectors in a sample of a polycrystalline Fe–3 wt.% Si alloy; strain increase 0.001.

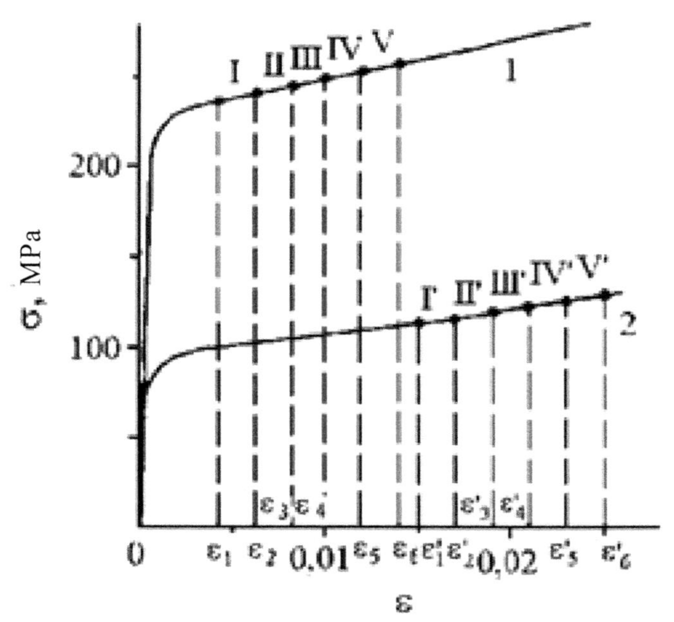

Fig. 2.3. Charts of loading samples of the Fe–3 wt.% Si alloy (1) and Al (2); I – V numbers indicate areas where specklograms were recorded.

The optical procedure for extracting information about the field of displacement vectors *r*(*x, y*) arising from the deformation of a flat sample (decoding of specklogram) is based on the fact that mutually shifted speckle systems form a diffraction grating. Its constant, which must be measured during decoding, is the displacement vector *r*. In the course of decoding, the circuit of which is shown in Fig. 2.1 b, the specklogram is scanned by an undiluted laser beam *1* with a diameter of ~1 mm. In this case, a system of alternating bright and dark Young fringes, separated by a distance *R*, appears on screen 3. They are the result of diffraction of parallel light beams on a diffraction grating – specklogram. 1970; Bychkov, Chuguy, 2011]

$$r\sin\varphi = 2n\frac{\lambda_{las}}{2} = n\lambda_l \qquad (2.1)$$

where φ is the angle between the optical axis and the direction under which the maximum is observed. The angle φ is small, so that sin φ ≈ tg φ ≈ *R L*, and with $n = 1$, the magnitude of the displacement vector with an increase in the total strain from ε_{tot} to $\varepsilon_{tot} + \delta\varepsilon_{tot}$ will be

$$|r| = \frac{\lambda_{las}L}{R}. \qquad (2.2)$$

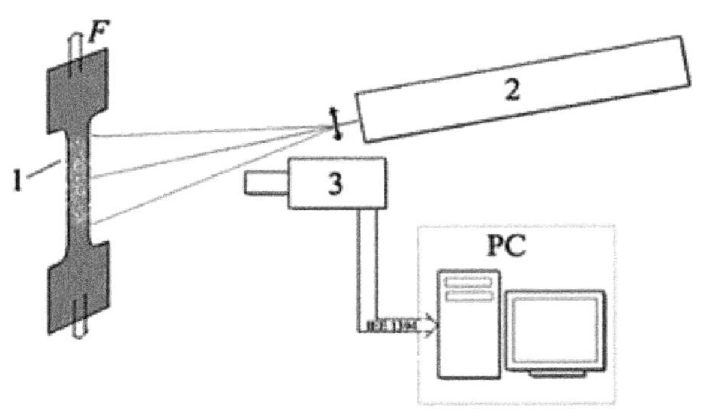

Fig. 2.4. Complex ALMEC-tv: 1 – sample; 2 – laser; 3 – video camera; PC– computer.

Table 2.1. Characteristics of the experimental complexes ALMEC and ALMEC-tv

Characteristic	Measurement unit	Recording mode	
		ALMEC	ALMEC– tv
		Photographic	Digital
Field of sight diameter	mm	~100	~100
Resolution	μm	1	2...3
Measurement accuracy		10^{-4}	10^{-4}
Recording mode		Frame by frame	Continuou
Recording frequency	s^{-1}	1	50
Operating mode		Subsequent data processing	Real time

For the setting used, λ_{las} = const and L = const, so to calculate the modulus of the displacement vector $|r|$, it is sufficient to measure the distance R on the screen. The direction of this vector is normal to the interference fringes. By successively scanning the obtained specklogram, it is possible to reconstruct the field of displacement vectors $r(x, y)$ for a flat sample, shown in Fig. 2.2. Usually in experiments $\delta\varepsilon_{\text{tot}} \approx 10^{-3}...2 \cdot 10^{-3}$. All measurements and calculations during decoding are performed automatically using specially developed programs.Using successive exposures in the deformation process it is possible, as shown in Fig. 2.3, record a series of strain specklograms along the entire plastic flow curve and obtain information on the temporal variations of the field r = (x, y, t) The number of such speklograms in the study of plastic materials can reach 200 or more. This procedure is rather laborious,

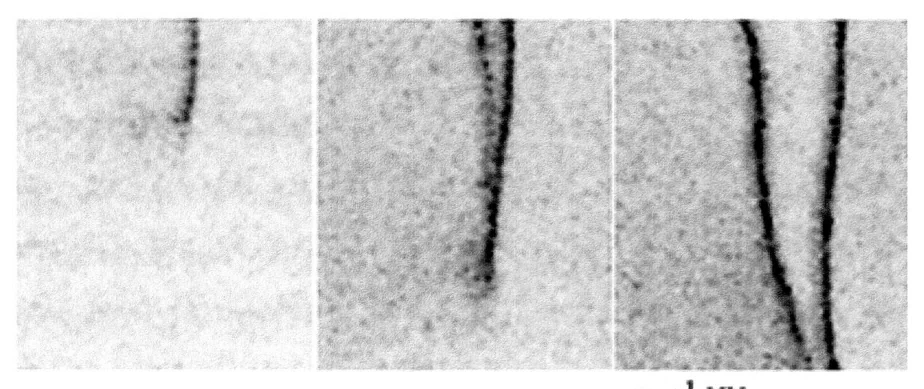

Fig. 2.5. Chernov–Lüders bands recorded at an interval of 7 from the nucleus growth stage across a mild steel specimen.

but it allows studying the kinetics of the deformation process) regardless of its duration.

In the developed method of recording and decoding sequential specklograms, it is not actually the absolute values of the displacement vector that are measured, but their increments with an increase in the total deformation from ε_{tot} to $\varepsilon_{tot} + \delta\varepsilon_{tot}$. If necessary, the total amount of displacement can easily be obtained by summing the increments of the displacement vector for each pair of exposures obtained during the recording process.

2.1.2. ALMEC-tv complex. Principle of operation

This complex was a further step in the development of methods for observing localized plasticity. Its distinguishing feature is the use of digital statistical speckle photography [Äbischer, Waldner, 1997; Sjödahl, 2001; Zuev, Gorbatenko, Pavlichev, 2010]. The block diagram of the complex is shown in Fig. 2.4.

When using this complex, the tensile-loaded sample is also illuminated with coherent light of a semiconductor laser with a wavelength of 635 nm and a power of 15 mW. The obtained images of the deformable sample with superimposed speckle patterns are recorded with a PixeLink PL-B781 digital video camera. A sequence of samples is formed for each point of the image, which characterizes the time course of its brightness, the variance and the expectation are calculated, which are used to display the deformation localization zones.

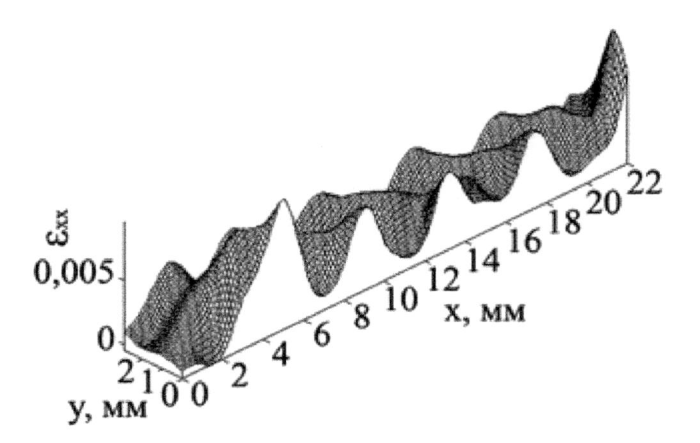

Fig. 2.6. To the calculation of the total deformation of the sample under tension.

Using this technique, it is possible *in situ* to register regions in which, at a given increase in the total elongation of the sample, the deformation of the material is localized. As an example, Fig. 2.5 shows photographs of the samples, where these areas look like narrow dark lines. They allocate places, where the plastic deformation is localized at the time of recording.

2.2. Patterns of localized plasticity

The use of the measuring complexes ALMEC and ALMEC-tv allows you to get pictures of the localization of the deformation and extract from them rich information about the kinetics of the development of this phenomenon in materials of different varieties. Localized spatial pictures, naturally or randomly changing in time and arising in environments of very different nature, are commonly called patterns in modern scientific literature [Cross, Hohenberg, 1993; Trubetskov et al., 2002; Scott, 2007; Haken, 2014]. This term was first introduced by Walter [1966], defining its meaning as follows: "The concept of a pattern implies any sequence of phenomena in time or any arrangement of objects in space that can be distinguished from another sequence or another arrangement, or compare with them." This concept will be used further to refer to observed patterns of localized deformation [Zuev, Danilov, Barannikova, 2001; Zuev et al., 2001], which need interpretation and qualitative and quantitative analysis.

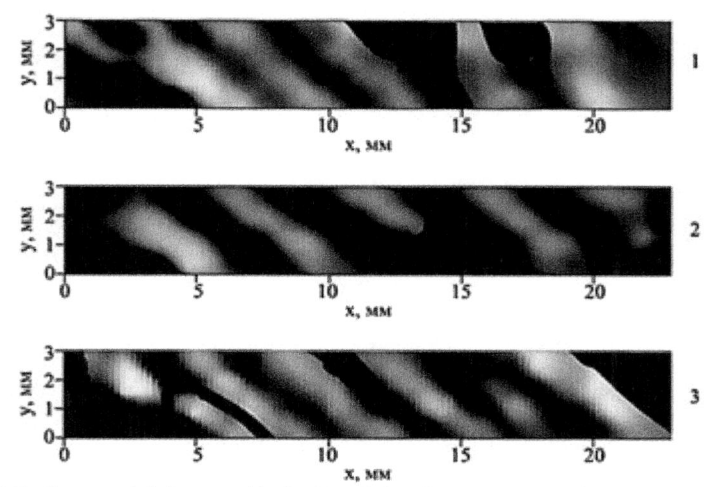

Fig. 2.7. Sequential frames (1, 2, 3) of the development of a localization pattern in a FeI single crystal at the stage of linear strain hardening

2.2.1. Patterns of localized plasticity and general deformation

To determine the quantitative relationship of the patterns obtained with a double-exposure speckle photograph with the total deformation of the sample, let us analyze the behaviour of the component of local elongation in the direction of the axis of stretching of the sample ε_{xx} in the sample deformed by $\delta\varepsilon \approx 0.002$. The calculation is made for the typical example of the distribution of ε_{xx}. From the presented data shown in Fig. 2.6, and it follows that plastic deformation is localized in several active zones of the sample, while other volumes of deformation material are practically not deformed. The same situation is characteristic of all the examples shown earlier.

If N is the number of deformation regions with the amplitude of deformation $\hat{\varepsilon}_{xx}^{(l)}$, l_i is the size of such a focus along the x axis and the average relative elongation within the focus $\langle\varepsilon_{xx}\rangle$, then the absolute elongation of the sample will be

$$\delta L \approx \sum_{i=1}^{N}\left(\hat{\varepsilon}_{xx}^{(i)} \cdot l_i\right) \approx N \cdot \langle\varepsilon_{xx}\rangle \cdot \bar{l} \tag{2.3}$$

In accordance with the data of Fig. 2.6, $N = 5$, the average length of the active focus is $\bar{l} \approx 3$ mm, the average deformation in the localization focus is $\langle\varepsilon_{xx}\rangle \approx \hat{\varepsilon}_{xx}^{i}/2 \approx 5\times10^{-3}$. Then formula (2.3) follows $d \approx L$ 0.075 mm, which corresponds to an increase of the total strain $\delta\varepsilon \approx \delta L/L \approx 0.002$, specified when recording the specklogram.

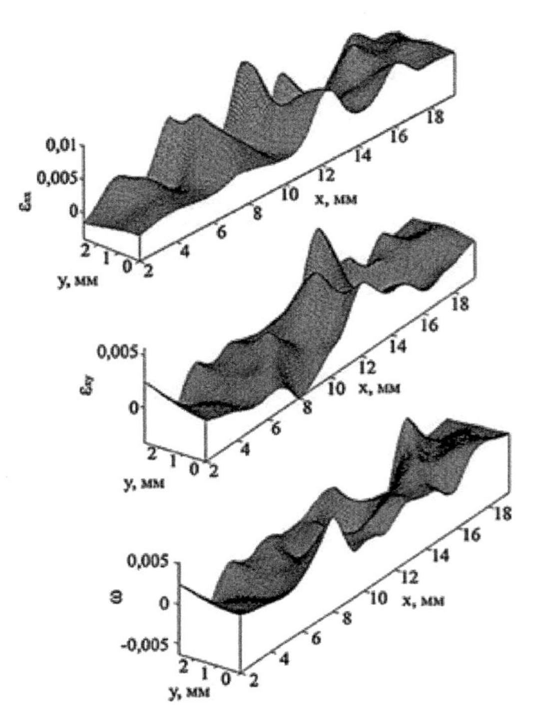

Fig. 2.8. A typical example of the distribution of the components of the plastic distortion tensor in a deformed FeI single crystal with an increase in the total strain of 0.002.

Thus, almost all plastic deformation of the sample during deformation accumulates in several thin ($l \ll L$) layers of the material – regions of localized plasticity. Such layers can be mobile or stationary, and their distribution along the sample, which evolves with an increase in total strain, is a pattern of localized plasticity. These results demonstrate the important role played by the localization of plastic flow in the deformation process.

2.2.2. Patterns of localized plasticity. Qualitative analysis

Specially developed software of the ALMEC and ALMEC-tv complexes allows different options for visualization of patterns of plastic flow localization. As an example, Fig. 2.7 shows a halftone pattern of successive instantaneous distributions of deformable and non-deformable layers of deformable material.

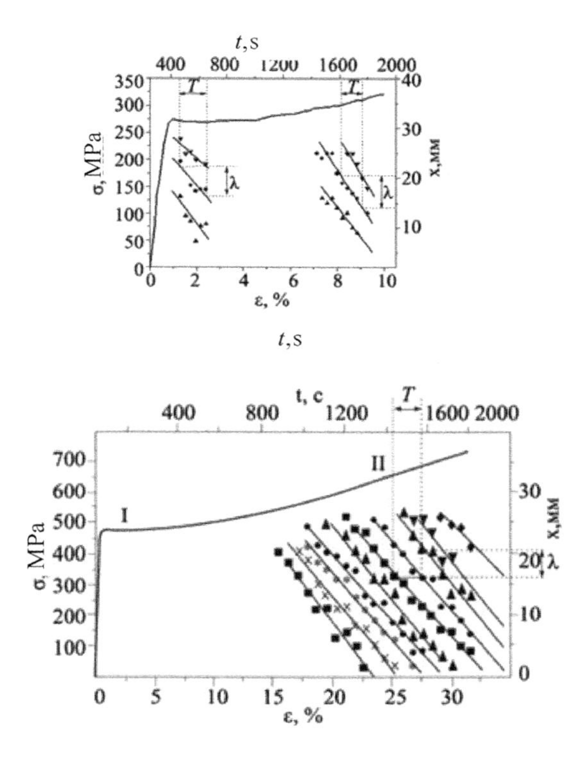

Fig. 2.9. On the methodology for determining the spatial and temporal periods of localized plastic deformation using *X–t* diagrams; (Fe_I single crystals at the stage of linear strain hardening).

The increase in tensile strain between individual frames was 0.002 for this case. The contrast for images is usually chosen in such a way that the dark areas correspond to the volumes that are currently deforming, and the light ones indicate passive volumes where there is no plastic flow at the time of observation. The color intensity is proportional to the amplitude of the local deformation. Such pictures allow us to estimate the spatial scale of plastic flow heterogeneity and to form an idea of the nature of its evolution.

2.2.3. *Patterns of localized plasticity. Quantitative analysis*

The source of quantitative information on the process of plastic flow is the data on the field of displacement vectors for a flat thin sample $r(x, y)$, which can be obtained from the original specklogram when they are decoded. The corresponding example of such a field is shown in Fig. 2.2.

The field gradient $r(x, y)$ is plastic distortion [DeVit, 1977; Kadic, Edelen, 1987]

$$\beta_{ij} = \nabla \boldsymbol{r}(x,y) = \begin{vmatrix} \varepsilon_{xx} & \varepsilon_{xy} \\ \varepsilon_{yx} & \varepsilon_{yy} \end{vmatrix} + \omega_z, \tag{2.4}$$

where around $\begin{vmatrix} \varepsilon_{xx} & \varepsilon_{xy} \\ \varepsilon_{yx} & \varepsilon_{yy} \end{vmatrix} \equiv \varepsilon_{ij}$ is the plastic strain tensor, and ω_z is the turn the z axis normal to the plane of the sample lying in the $x0y$ plane and stretchable in the x direction. The example in Fig. 2.8 shows the result of calculating the components ε_{xx}, ε_{xy} and ω_z of the tensor (2.4).

The longitudinal $u = f \cos \phi$ and transverse $v = r \sin \varphi$ components of the displacement vector r and the angle ϕ between the vector r and the direction of the stretching axis x, which are necessary for the calculation, are determined when decoding the speclogram. The components of the plastic distortion tensor are local elongation $\varepsilon_{xx} = \partial u/\partial x$, local narrowing $\varepsilon_{yy} = \partial v/\partial y$, shift $\varepsilon_{xy} = \varepsilon_{yx} = 1/2(\partial v/\partial x + \partial u/\partial y)$ and rotation around the z axis $\omega_z = 1/2(\partial v/\partial x - \partial u/\partial y)$ are calculated using a special computational program. The values obtained in the calculation are the increments of local deformations (or turns), and not their integral values from the beginning of the loading process.

2.2.4. Kinetics of development of patterns

Quantitative information on the evolutionary characteristics of localized plastic flow is the results of an assessment of the spatial and temporal scales of patterns of localization of plastic flow, detected using the ALMEC and ALMEC-tv installations. For such an assessment, the dependences of the position of the localized plasticity X on the strain $X(\varepsilon)$ were used, which are compared with the $\sigma(\varepsilon)$ curves. With active loading, $\varepsilon \sim t$, so it is easy to obtain

Table 2.2. The materials studied (the composition of the alloys is indicated in wt.%)

Composition		
Pure metals	BCC	Nb, Ta, Mo
	FCC	Al, Cu, Ni, Pb
	HCP	Zn, Cd, Mg, Co, Ti, Hf
	Tetr.	Sn, In
Alloys based on	α-Fe	Fe–3%Si, Fe–0,1%C

the dependences $X(t)$, which are called X–t diagrams. Their example is presented in Fig. 2.9.

It is clear that the horizontal sections (X = const) determine the time (T), and the vertical (t = const) the spatial (periods (scales) of the pattern evolution process. One can also use the frequency $\omega = 2\pi/T$ and the wave number $k = 2\pi/\lambda$ process of evolution.The speed of movement of deformation regions is defined as $V = \lambda/T = \omega/k$.

The numerical values of these values are close for all materials studied and are: time period $T \approx 10^2...10^3$ s; spatial period $\lambda = 10^{-2}$ m; the speed of movement of the regions $V = \lambda/T \approx 10^{-5}...10^{-4}$ m/s.

This indicates a remarkable commonality of the observed patterns for all studied metals and alloys. The values of T, l, and also the velocity V are convenient characteristics for a general quantitative description of the phenomenon of the macroscopic localization of deformation.

2.3. On the choice of materials for research

The choice of materials for the study was dictated by the desire to identify the general patterns of development of localized plastic flow characteristic of the deformation process in general, regardless of the structure and properties of specific metals and alloys. For this reason, plastic deformation has been studied for a wide range of metals and alloys belonging to different crystal classes and in different structural states, as well as deforming due to different microscopic mechanisms – dislocation slip, twinning, and phase transformation deformation. Some non-metallic materials were also studied. A general summary of the materials is given in Table 2.2.

When working with single-crystal samples, the staging and shape of plastic flow curves was varied by choosing the orientation of the stretching axis [Berner, Kronmüller, 1969]. These orientations are not indicated in the corresponding figures, since they do not play a decisive role when discussing experimental results.

2.4. Stage of plastic deformation and localization patterns

A characteristic feature of the plastic flow process is its multi-stage character, that is, the existence of the $\sigma(\varepsilon)$ dependences, which are characterized by different form of the $\theta(\varepsilon)$ dependence [Seeger, 1960; Mughrabi, 1983; Argon, 2008; Messerschmdt, 2010]. One of the objectives of this study was to find a possible connection between the strain hardening law $\theta(\varepsilon)$ acting at each stage of the process and the pattern of localized plasticity observed at the same stage. Such a statement of the problem is quite natural, since the multistage process of the flow determines the structure of the models of work hardening materials, aimed at explaining the nature of the stages of the process. Theories of strain hardening are usually constructed in such a way as to explain the nature of the phenomena responsible for each stage of the deformation process (see, for example, Seeger, [1960]; Berner, Kronmüller, [1969]; Kuhlmann-Wilsdorf, 2002).

2.4.1. Selection of stages of the plastic flow curve

The analysis of the process of plastic flow begins with the selection on the flow curve of the $\sigma(\varepsilon)$ sections characterized by some constant value and corresponding to certain laws of strain hardening $\theta(\varepsilon)$. Selection can be done in different ways. It is convenient, for example, to approximate the plastic flow curve $\sigma(\varepsilon)$ by the empirical equation of Ludwiq [Hill, 1956; Shiratori, Miyoshi, Matsushita, 1986]

$$\sigma(\varepsilon) = \sigma_0 + Q \cdot \varepsilon^n \tag{2.5}$$

in which σ and ε are the current values of stress and strain; Q is the hardening rate, and σ_0 is the yield strength. Then the stages of the flow curve can be characterized by the parameter of parabolicity n in equation (2.5). Processing the curves $\sigma(\varepsilon)$ using equation (2.5) showed that the value $n = \frac{\ln[(\sigma - \sigma_0)/Q]}{\ln \varepsilon}$ changes discretely along the curve $\sigma(\varepsilon)$. This provides a confident selection of the stages of the process of plastic flow on curves $\sigma(\varepsilon)$ according to the following features:

– $n \approx 0$ at the flow area,

– $n \approx 1$ at the stages of light slip and linear strain hardening,

– $n \approx 1/2$ at the stage of parabolic hardening,

– $0 \leq n \leq 1/2$ at the pre-failure stage.

The applicability of equation (2.5) to isolate the stages of the deformation process was verified by comparing with the results of the construction of dependences of the strain hardening coefficient

Table 2.3. Observation of strain hardening stages

Metal, alloy, deformation mechanism	Single or polycrystal	Observed stage of strain hardening
Cu	single crystal	easy slip, linear hardening, parabolic hardening
Ni	single crystal	easy slip, linear hardening, parabolic hardening
Al	polycrystal	linear hardening, parabolic hardening
Fe_I– 16%Cr –12% Ni–2%Mo	single crystal	easy slip, linear hardening, parabolic hardening
Fe_{II}– 13%Mn; twinning	single crystal	easy slip, linear hardening,
Cu – 10%Ni – 6%Sn	single crystal	easy slip, linear hardening, parabolic hardening
Ni_3Mn	polycrystal	yield plateau, parabolic hardening
Mg – 2%Mn	polycrystal	linear hardening, parabolic hardening
Zn	single crystal	easy slip, linear hardening,
Zr – 1%Nb	polycrystal	linear hardening, parabolic hardening
Fe – 0.08%C	polycrystal	yield plateau parabolic hardening
Fe – 3%Si	single crystal	linear hardening, parabolic hardening
Ta, Hf	polycrystal	linear hardening, parabolic hardening
NiTi (equiatomic composition) phase transformation deformation	single crystal	
Dislocation slip mechanism operated in other cases		

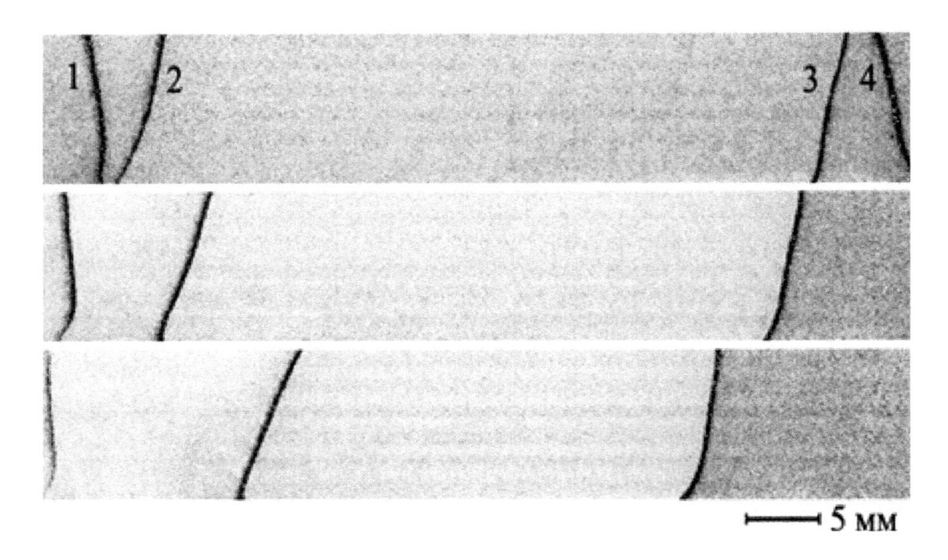

Fig. 2.10. Counter-movement of fronts (2 and 3) of the Chernov-Lüders bands.

on the strain $\theta(\varepsilon)$ or on the stress $\theta(\sigma)$(that are commonly used for these purposes [Malygin, 2001; Pelleg, 2013]. It turned out that the method used yields almost the same results, but it is attractive with significantly less labor input.

Table 2.3 presents the data on the materials during the deformation of which the listed stages of strain hardening were observed.

2.4.2. The yield plateau stage

At the site of yield, materials are deformed at a constant stress, that is, $n \approx 0$ and $\theta = 0$. As shown by numerous experiments, plastic deformation in such materials develops by expanding one or several Chernov–Lüders bands, whose fronts separate plastically deformed and elastic regions [Thomas, 1964; Mac Lean, 1965; McClintock, Argon, 1970; Honeycomb, 1972; Pelleg, 2013]. Several such bands and fronts in a sample can simultaneously arise and coexist.

In our experiments [Gorbatenko, Danilov, Zuev, 2017; Zuev, 2017a, b], made with the help of the ALMEC-tv complex, we were able to establish that the fronts of one band move in opposite directions along the axis of the sample with the same speeds. As one of the fronts approaches the capture of a car, its speed drops to zero, and the speed of the second one is approximately doubled. With the simultaneous initiation of two bands (Fig. 2.10), at first all

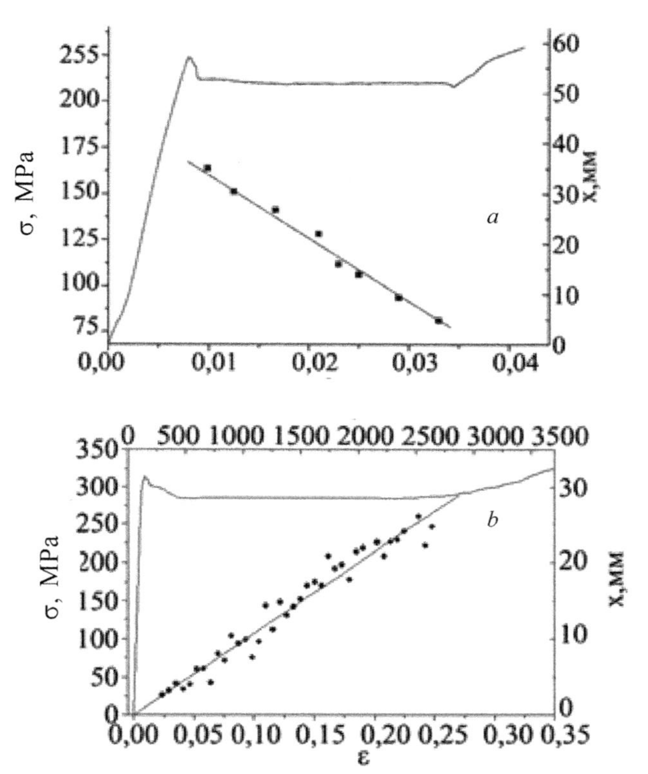

Fig. 2.11. $X - t$ diagrams and plastic flow curves at the yield plateau: mild steel (a) and Fe_{II} alloy (b),

four fronts move with almost identical speeds. When the fronts 1 and 4, approaching the grips, stop, the remaining fronts 2 and 3 move towards each other with doubled speed compared to the initial speed. In the general case, for the fronts of the Chernov-Lüders band in a deformable sample, the rule

$$\sum_{i=1}^{N}\left|V_f^{(i)}\right| = \tilde{V}_f = \text{const}, \tag{2.6}$$

in which $\left|V_f^{(i)}\right|$ – is the velocity of the i-th front, N is the number of moving fronts, and $\tilde{V}_f \approx 1.6 \cdot 10^{-4}$ m/s. Condition (2.6) ensures the constancy of the growth rate of the area of the plastically deformed zone in the sample.The meeting of the moving fronts in the development in the sample of two Chernov-Luders bands can be carried out according to two scenarios. When implementing the first one, there is a penetration of fronts in the region of adjacent

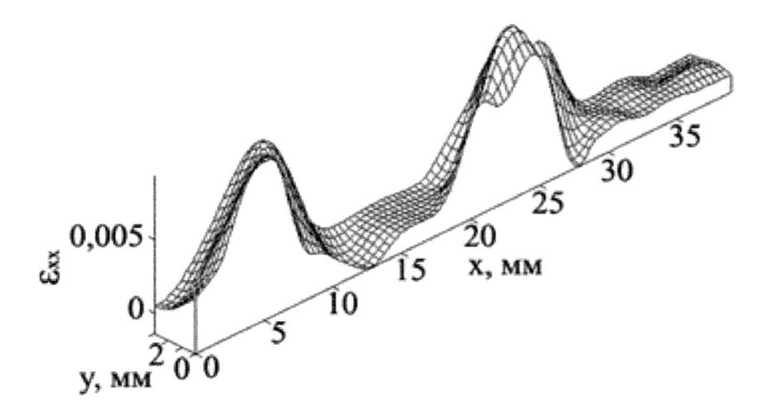

Fig. 2.12. Easy slip localization pattern.

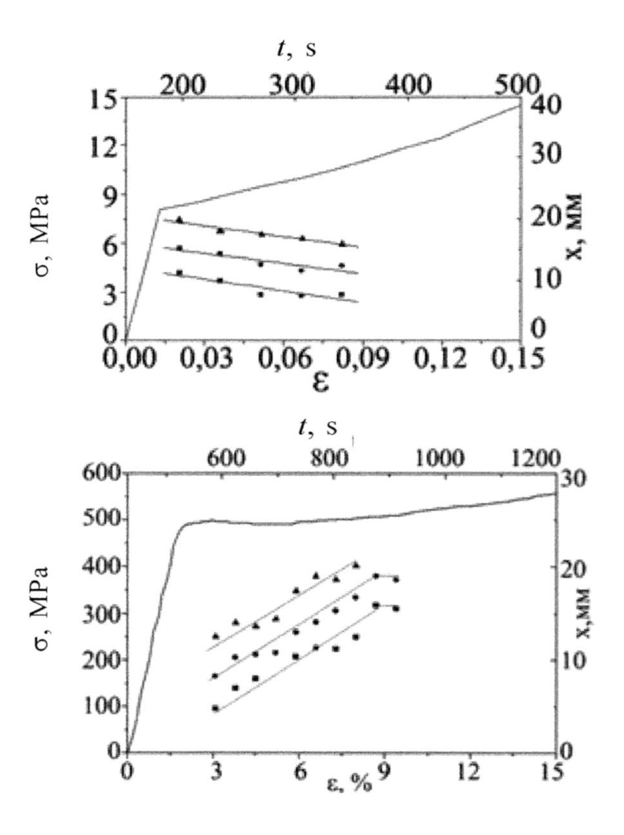

Fig. 2.13. Plastic flow curves and X – t diagrams for the easy slip stage: Cu (a) single crystals; FeII (twinning strain) (b).

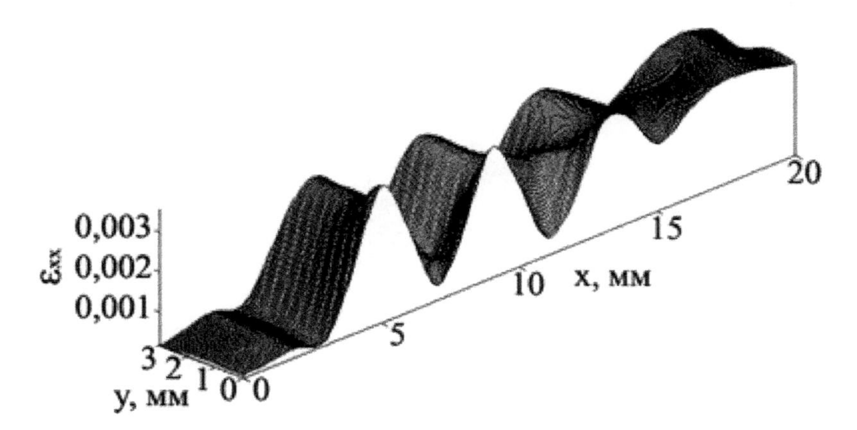

Fig. 2.14. Example of localization at the stage of linear strain hardening in a Fe_I single crystal: strain increase 0.002.

strips, that is, plastic deformation, which occurred before in one strip, continues for some time in a deformed region related to another. In the second scenario, the zone between the fronts is fragmented with the formation of secondary fronts connecting the primary ones. In any case, at the meeting of the fronts, the stage of the yield plateau ends and further deformation continues with strain hardening, that is, the fronts of the Chernov-Lüders strips cannot pass through each other. Typical X–t diagrams for cases of the propagation of the fronts of the Chernov–Lüders band at the yield plateau area are shown in Fig. 2.11.

The pattern of localized plasticity corresponding to the yield plateau is a deformation front moving along the sample at a constant speed. It separates the areas in the elastic and plastically deformed states, that is, at the front, the material passes from the elastic to the plastically deformed state. When the front moves, the area of elastically deformed areas decreases, and the area of plastically deformed areas increases.

2.4.3. Stages of easy slip and linear hardening

At the stages of easy sliding and linear strain hardening, deforming stresses are proportional to the deformation $\sigma \sim \varepsilon$, $n \approx 1$ and θ = const. At the same time, the strain hardening coefficients at these stages θ_{eg} and θ_{lwh}, respectively, are constant, but differ in magnitude, so that usually $\theta_{lwh}/\theta_{eg} \approx 30$ [Seeger, 1960; Berner, Kronmüller, 1969; Honeycomb, 1972].

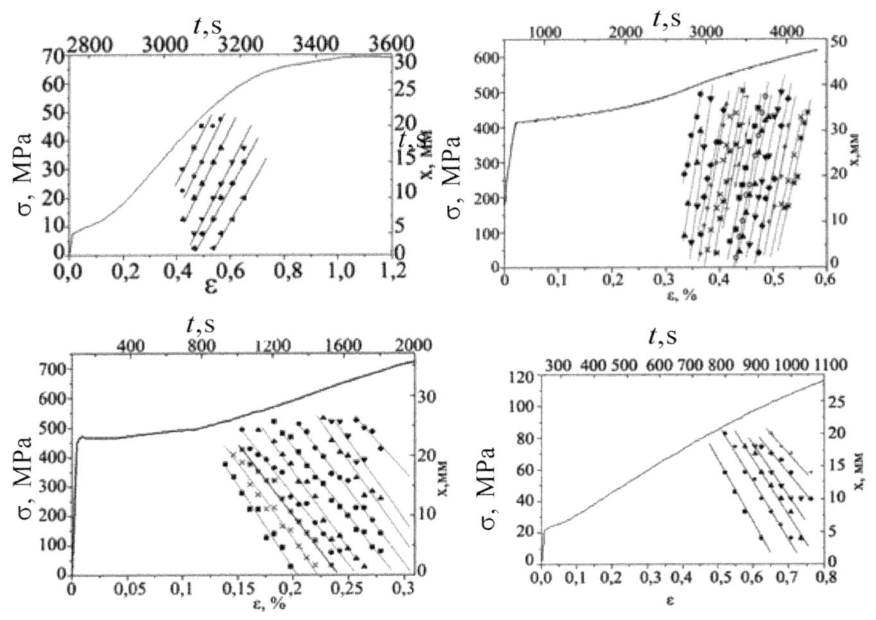

Fig. 2.15. Plastic flow curves and *X–t* diagrams for the linear hardening stage: Cu (a) single crystals; Fe_{II} (*b*); Fe_{II} (*c*); Ni (*d*).

In the single-slip monocrystals, the patterns of localized deformation are a combination of two or three deformation zones disposed at equal distances and simultaneously moving along the sample with a constant velocity. A typical for such cases pattern of localization of plastic deformation, observed in a Fe_{II} single crystal, is shown in Fig. 2.12.

Figures 2.13, a, b (lines 2–4) shows the combined diagrams of plastic flow and *X–t* diagrams of such zones for copper single crystals and an Fe_{II} alloy, oriented for easy slip. From the *X–t* diagrams constructed, it follows that at the stages of easy slip along the samples, three sources of plastic flow move synchronously along the samples. The distance between the regions does not change during their movement and is ~5 mm. The velocity of movement of the deformation zones, determined from the slope of the $X - t$ diagrams presented, reached ~ 4×10^{-5} m/s, which is slightly lower than the velocity of movement of the front of the Chernov–Lüders band.

At the stage of linear strain hardening of single and polycrystals, the observed localization patterns have the same structure. However, at this stage, in the sample along the stretching axis, a larger number

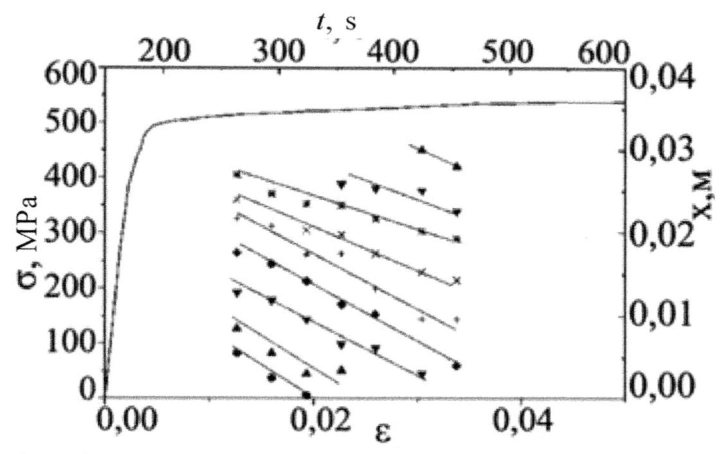

Fig. 2.16. Plastic flow curve and $X-t$ diagram for the stage of linear strain hardening of a single crystal of Fe-3 wt.% Si alloy.

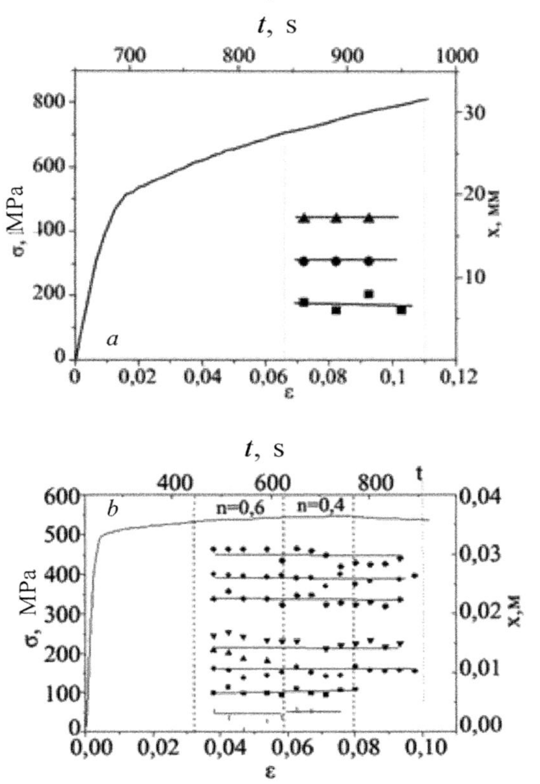

Fig. 2.17. Plastic flow curves and $X-t$ diagrams for the parabolic hardening stage: single crystals: Fe_1 (*a*); Fe–3 wt.% Si (*b*).

of localization regions synchronously move, which is usually 5–15, as can be seen in Fig. 2.14.

The use of *X–t* diagrams (Fig. 2.15) showed that in the cases studied (the stage of linear strain hardening), the patterns of localized plasticity are characterized by constant distances between the localization centers (spatial period) ~3 ... 8 mm and constant speed of movement of the centres ~ $10^{-5}...10^{-4}$ m/s.

It was also established that at this stage the speed of movement of localized plasticity centres is inversely proportional to the strain hardening coefficient of the material under study, that is, condition $V \sim \theta^{-1}$. The importance of this fact will be discussed further (Chapter 3) when choosing equations describing the kinetics of the processes of development of localized plastic flow in deformable media.

To verify the generality of the results obtained, the localization of plastic flow was investigated at the stage of linear strain hardening of the bcc single crystal of an Fe–3 wt.% Si alloy. The corresponding results in Fig. 2.16 show that in this case, during the entire stretching process, the deformation is distributed non-uniformly as in fcc single crystals and the spatial period of the localization regions is ~ 5 mm. The speed of the moving lozalization regions is $-5.8 \cdot 10^{-5}$ m/s. Thus, in the bcc alloy, the patterns of development of localized plasticity are similar to those characteristic of fcc metals and alloys.

Thus, the formation of an equidistant system of synchronously moving regions of localized plasticity is a common property of the materials studied, the deformation of which leads to a linear relationship between the deforming stress and strain. This is true for both mono- and polycrystalline states.

2.4.4. Stage of parabolic hardening

At this stage of hardening, the dependence of the flow stress on the strain is $\sigma \sim \varepsilon^{1/2}$, and $\theta \sim \varepsilon^{-1/2}$. A similar dependence is characteristic of many single- and polycrystalline materials. At these stages, the pattern of localization of plastic flow had the appearance of a stationary system of deformation regions, that is, along the length of the samples at a distance of ~8 mm from each other there were three stationary deformation points, as shown in Fig. 2.17 *a*.

Additionally, single-crystal samples of an Fe–3 wt.% Si alloy were investigated, in which, during deformation, a parabolic stage of deformation hardening is also detected. The macrolocalization

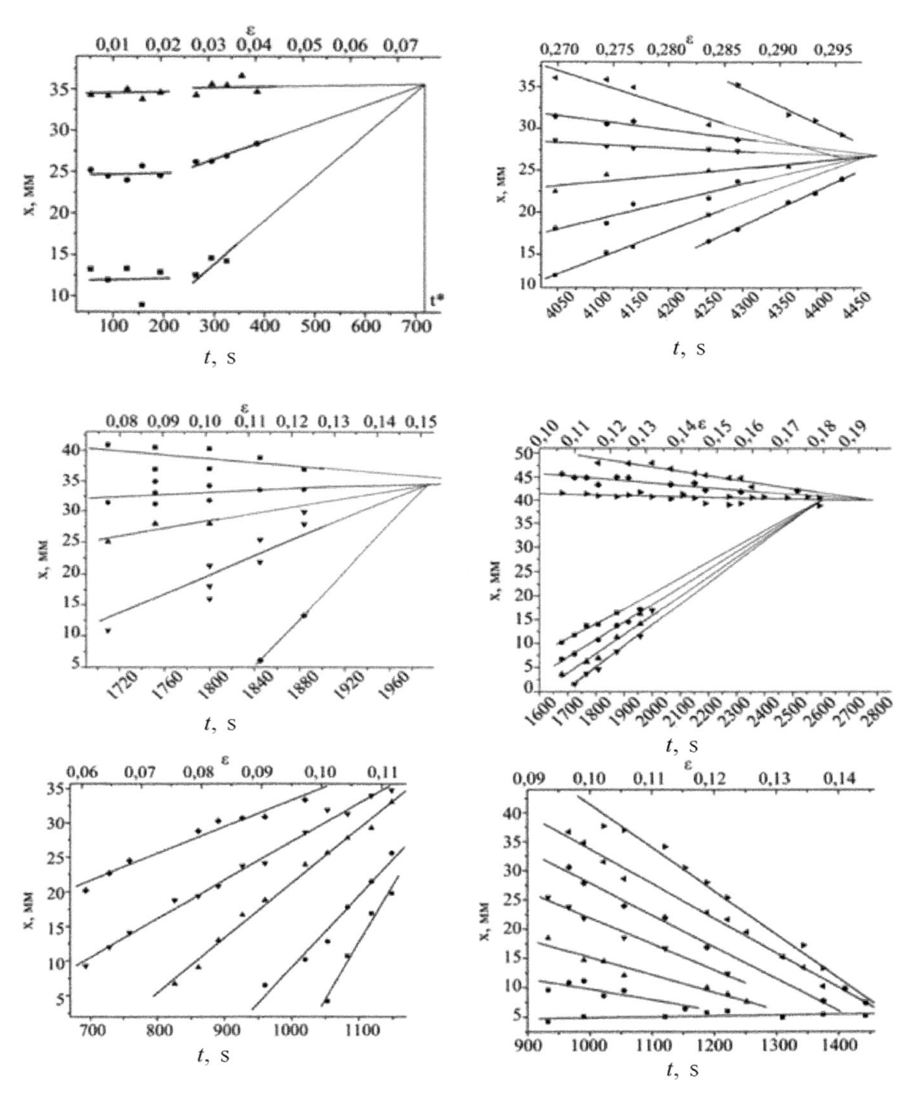

Fig. 2.18. *X–t* diagrams for the pre-fracture stage: fine crystalline Al (*a*); coarse crystalline Al (*b*); Al–4 wt.% Cu (*c*); Fe–3 wt.% Si (*d*); Mg–2 wt.% Mn (*e*); Zr – 2.5 wt.% Nb (*f*).

patterns in this alloy are shown in Fig. 2.17, the combination of fixed equidistant deformation sites with approximately the same amplitude.

2.4.5. Pre-fracture stage

At the next (after the parabolic hardening stage) the pre-fracture

Table 2.4. Comparison of experimentally recorded and calculated coordinates and time of destruction of samples

	Al	Ti	Fe–3%Si	V–2.3%Zr–0.4%C
$\hat{x}_{exp}/\hat{x}_{calc}$	1.1	1.1	1.0	1.0
t^*_{exp}/t^*_{calc}	1.0	1.1	1.0	1.0

stage, the strains become mobile again, but the nature of the process at this stage is more complicated than that observed, for example, at the stage of linear deformation hardening. $X(t)$ are rectilinear and at extrapolation converge at the point with coordinates $\hat{x}^*$ and t^*, forming beams of straight lines. The speeds of movement of individual regions remain constant, that is, they are mutually consistent from the very beginning of the pre-destruction stage. Examples of X–t diagrams of localized deformation at the pre-fracture stages for a number of materials studied are shown in Fig. 2.18.

Comparing these data with plastic flow curves shows that such patterns of localized deformation are characteristic of the stage preceding the onset of the fall of conditional stresses on the $\sigma(\varepsilon)$ diagram, that is, until the Drucker postulate of the plasticity theory on the positiveness of external forces is satisfied [Goodier, Hodge, 1960; Struzhanov, Mironov, 1995; Han Chin-Wu, 2005]. In other words, the dependence shown in Fig. 2.18 are valid for the stress interval $\sigma_* < \sigma < \sigma_B$ where σ_* is the stress at the end of the parabolic work hardening stage, and σ_B is the ultimate strength.

Based on the analysis of these data, the following important patterns have been established, which are characteristic of the development of patterns of localized plastic flow at the pre-fracture stage at $n < ½$:

- in the coordinates X–t, the regions of a localized plastic flow move along straight lines forming a beam having a centre;
- from the beginning of the pre-fracture stage, the localized deformation regions acquire constant vlocities, which are different for each regions.

The moving regions of localized plasticity that appear at the pre-fracture stage, which form the pattern characteristic of the pre-fracture stage, are not related to the fixed ones that previously existed at the parabolic hardening stage. New regions of localized plastic flow can arise both on one side of the place of future fracture, and on either side of it (Fig. 2.18, d).

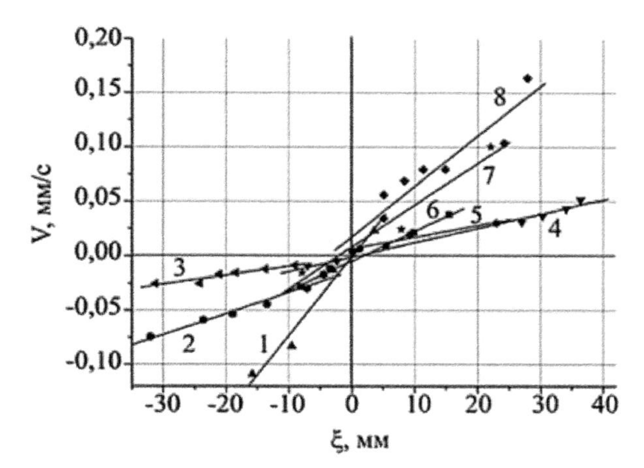

Fig. 2.19. The dependences $V(\hat{x})$ in the alloy V–2.3% Zr–0.4% C (1); in the alloy Zr–2.5% Nb (2); in Ti (3); in Fe–3% Si (4); in fine crystalline Al (5); in crystalline Al (6); in the Al–4.5% Cu–0.5% Mn–0.5% Mg alloy (7); in Mg (8).

The fact that at the pre-fracture stage, the dependences $X(t)$ for different metals and alloys form beams of straight lines with the coordinates of the centres $\hat{x}^*$ and t^* means that from the beginning of the pre-fracture stage, the speeds of the localized deformation sites is synchronized automatically ensuring the simultaneous arrival at the centre. For the formation of beams of straight lines $X(t)$ it is necessary that the speeds of movement of the regions depend linearly on the coordinates of the place of their origin $\hat{x}$, that is, the relation

$$V_{aw}\left(\hat{x}\right)=\alpha_0+\alpha\hat{x} \tag{2.7}$$

in which α and α_0 are empirical constants, and the x coordinate is measured from the fixed source of localization, is satisfied. Linear dependences $V(\hat{x})$ for the materials studied are shown in Fig. 2.19. The linear nature of the dependence (2.7) allows us to find the point of intersection of the lines on the corresponding graphs by extrapolating the first values after the process begins. This point obviously corresponds to the conditions $\hat{x} = \alpha_0/\alpha$ and $t^* = t_0 + 1/\alpha$ (t_0 is the time since the beginning of the pre-fracture stage). Table 2.4 compares the coordinates of the centres of the beams thus calculated with the experimentally observed ones. Obviously there is a good agreement between the calculated and experimentally observed values. Thus, it can be assumed that at the pre-fracture stage events develop according to a single scenario, which is ensured by the

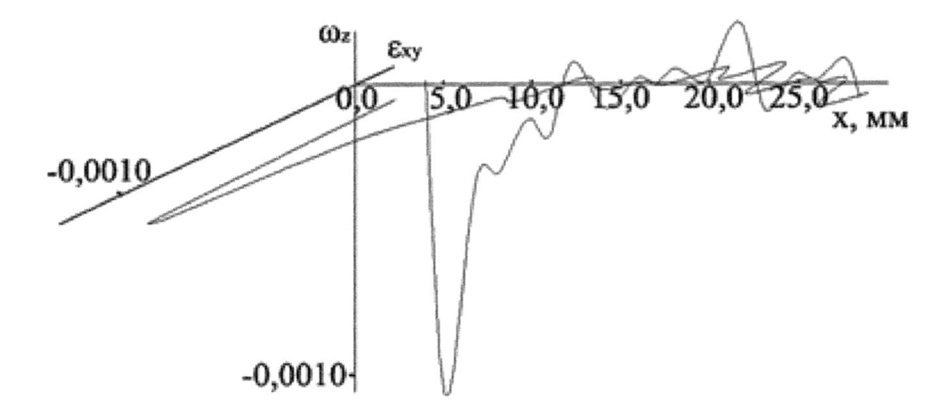

Fig. 2.20. The maxima ε_{xy} and ω_{wz} of the components of the plastic distortion tensor in a single crystal of Fe–3 wt.% Si alloy.

automatic fulfillment of relation (2.7) for the velocity of regions in the stretched sample as a function of their nucleation sites $\hat{x}$ for $n < ½$. This makes it possible to determine fracture parameters with acceptable accuracy for predicting fracture [Collacott, 1989]. This fact can be used to predict the moment and place of destruction during mechanical testing or in the operation of a product [Zuev et al., 2000; Breeders, Zuev, Kotrekhov, 2014].

Analysis of the kinetics of regions of localized plasticity with increasing strain showed that at the moment of the origin of a viscous crack, they all merge at the place of its formation, almost simultaneously coming to this section of the sample. Consequently, the place of destruction and the lifetime of the sample before destruction are determined by the processes occurring at earlier stages of plastic flow. The sharp maximum of strain localization, which serves as a precursor to the formation of a source of destruction, appears at the site of the future female at the end of the parabolic hardening curve at $n \approx ½$, when all the regions are fixed. After its formation, the other localization regions are consistently moving along the axis of stretching with speeds determined by equation (2.7). This pattern allows all regions of localized plasticity to arrive in the area of destruction at the same time. In other words, the process of fracture is prepared by processes of plastic flow taking place. This point of view is consistent with the views of Stepanov [1953], according to which plasticity and destruction are causally related, and destruction is the final stage of plastic deformation.

The confluence zone of plasticity centres moving at the pre-fracture stage has a complex structure. In the single crystal of the Fe–3 wt.% Si alloy, it was possible to observe the formation of a high-amplitude maximum of localized deformation at the end of the parabolic hardening stage with a parabolicity index of $n < 0.5$, as shown in Fig. 2.20.

As the above data show, the formation of the neck and the transition from plastic flow to viscous fracture when $n < ½$ is realized by coordinated movement of localized flow regions in such a mode that the distance between them gradually and regularly decreases and, finally, the regions are merge. The decrease in the distance between the regions of localized plasticity distinguishes the nature of their movement in this case from the character of movement at a constant speed at the stage of linear strain hardening, when the distance between the regions is constant. The described patterns can be considered common to plastic materials.

The final stage of the process of plastic deformation determines the place of formation of the neck. At this stage, the processes of plastic flow freeze in the entire volume of the sample, except for the part immediately adjacent to the site of origin of the neck. The consistency of the velocities of the regions at this stage is a sign of the process of *self-organization*, which is manifested in the synchronization of the movement of the regions of local plastic flow along the sample. Of all the regions, over time, only one retains activity, the position of which already at its origin corresponds to the place of formation of a macroscopic neck in the future and, accordingly, viscous destruction. Appearing at the stage of parabolic strain hardening at $n \approx 1/2$, such a region remains almost immobile

Table 2.5. The compliance rule of stages of strain hardening and patterns of localized plasticity

Strain hardening stages	Localized plasticity patterns
Yield plateau	Movement of fronts separating elastically and plastically deformed regions
Esy slip	A set of 2–3 moving regions
Linear strain hardening	A set of 5–10 moving regions
Parabolic strain hardening	A set of stationary regions
Pre-fracture	Merger of regions by necking

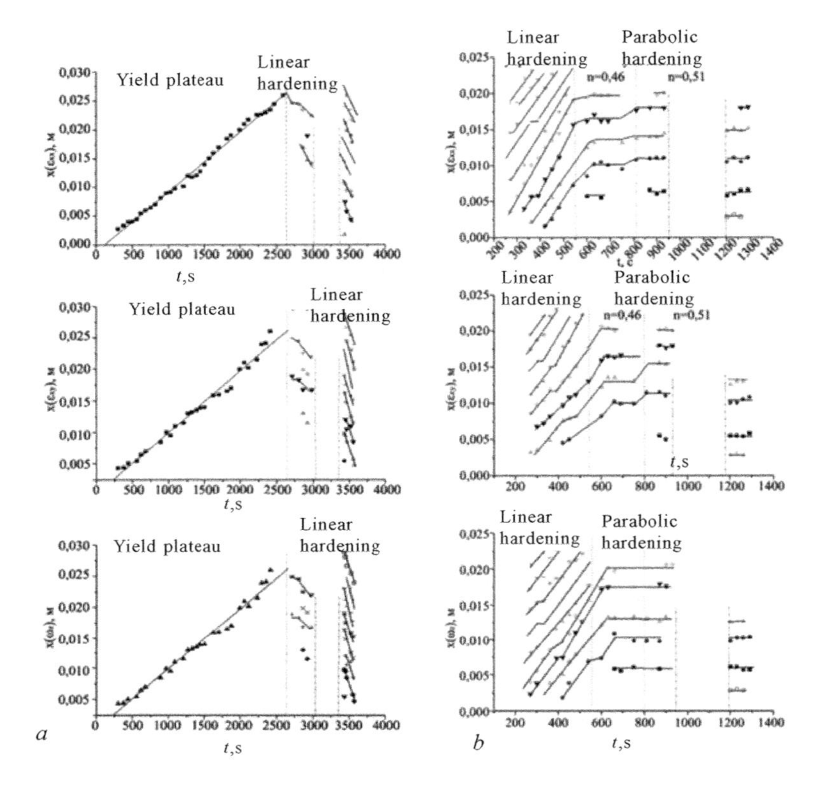

Fig. 2.21. The kinetics of regions of localized deformation in accordance with the staging of the plastic flow curves of single crystals: Fe_{II} (*a*); Fe_{I} (*b*).

until destruction, but the deformation in it gradually grows as the activity of the plastic deformation process in other regions decays.

2.5. The evolution of localization patterns during interstage transitions

Relationships of changing the pattern of localized plasticity in the interstage transition are very important for understanding the nature of autowaves of localized deformation. The technique of the *X–t* diagrams makes it possible to analyze the kinetics of the process of such a shift.

2.5.1. Transition patterns

The results of the study of such a shift are presented in Fig. 2.21,

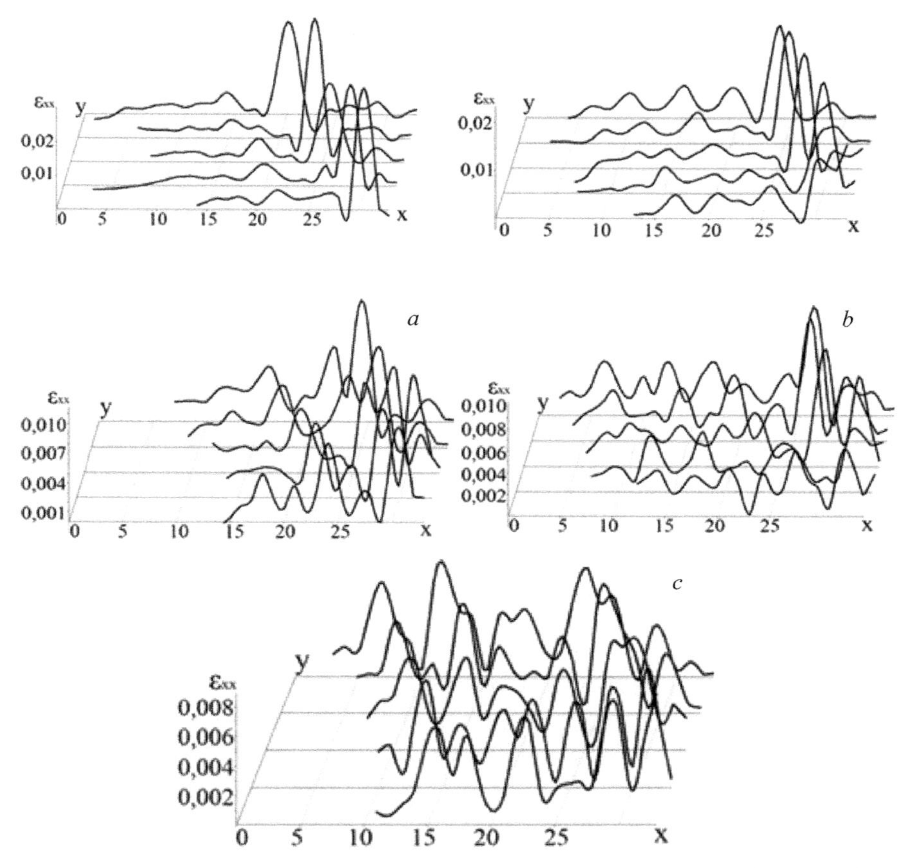

Fig. 2.22. The restructuring of the autowave pattern in the transition from the easy slip stage to the stage of linear strain hardening during the deformation of an Fe_{II} single crystal. Distributions of components ε_{xx} on patterns 0→*d* registered in 10 seconds.

a, b, where *X–t* diagrams are shown characterizing the evolution of the components of the plastic distortion tensor (ε_{xx}, ε_{xy} ω_z) for a Fe_{II} single crystal during the transition from the yield point to the linear strain hardening stage and for the Fe_I single crystal during the transition from the stage linear to the stage of parabolic strain hardening.

It is clear from this figure that each stage of plastic flow is characterized by a pattern of localization of plastic deformation that is quite definite in form, and the components of the plastic distortion tensor change in the course of the corresponding interstage transition in a completely regular way.

Figure 2.22 demonstrates successive (a → e) stages of the process of restructuring a pattern that represents one focus of

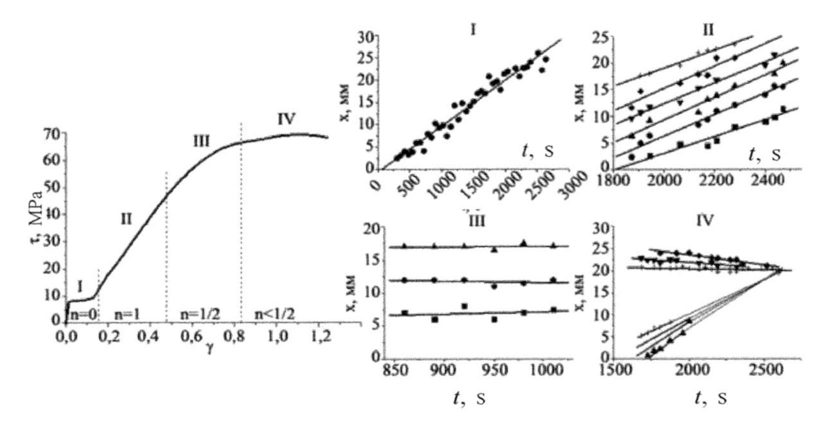

Fig. 2.23. To the formulation of the **Rules of Compliance** for the correspondence of the stages of strain hardening and the types of $X - t$ diagrams.

localized plasticity (easy slip stage, Fig. 2.22 a) to a sequence of synchronously moving lesions forming a pattern characteristic of the linear strain hardening stage (Fig. 2.22 d). It can be seen that such a transition is realized through the destruction of the original pattern, the randomization of the pattern of localization of the plastic flow and the formation of a new pattern (four regions) from deformation chaos.

2.5.2. Compliance Rule

It is clear that each stage of strain hardening in poly- and single crystals of the studied metals and alloys corresponds to a certain pattern of localized plasticity. It can be assumed that the forms of patterns and the characteristics of the stages of strain hardening are related by the **Compliance Rule**, which is illustrated in Fig. 2.23, schematically summarizing the patterns described above. This Rule, verified for all materials currently studied, is formulated in a generalized form in Table 2.5.

Thus, the Compliance Rule is satisfied for all materials, regardless of their composition and structure, as well as the mechanism of plastic flow. Quite wide variations in the structure and microstructure of materials entail only quantitative changes in the localization patterns, without affecting its main characteristics. The observed forms of such a correlation are exhausted by the regularities listed above, and their number coincides with the number of stages of the plastic flow curve, indicated in Fig. 2.23 in roman numerals.

3

Plastic flow as an autowave process

The experimental data presented in Chapter 2 prove the validity of the opinion that when stretching the initially structurally homogeneous sample at a constant speed, its plastic deformation proceeds in a localized manner from the yield point to fracture. In this interval, which in plastic materials can reach tens of percent, the patterns of localized plastic flow are generated and evolve, that is, the space-time heterogeneity (structure) of the deformable material is formed and develops. According to Seeger and Frank [Seeger, Frank, 1987], inhomogeneous defect structures arising in deformable media should be considered as the result of spontaneous ordering of the deformable medium.

This idea in recent decades has received a powerful impetus to development, thanks to the penetration of ideas and methods of synergetics into the physics of plasticity [Nicolis and Prigogine, 1990; Olemskoi, Khomenko, 2001; Olemskoi, Katznelson, 2003; Olemskoi, Kharchenko, 2007]. The fruitfulness of their application to the problem of plastic flow has been demonstrated by many authors [Ivanova, 1989; Zasimchuk, 1989; Naimark, Ladygin, 1993; Ivanova et al., 1994; Naimark, 1998]. In the 1980s, an approach began to take shape at plastic deformation as a specific process of self-organization of the deformation structure and its subsequent evolution [Vladimirov, 1987; Seeger, Frank, 1987; Nicolis and Prigogine, 1990; Olemskoi, Sklyar, 1992; Zuev, 1994; Naimark, 2003; Zuev, 2001, 2007, 2012].

3.1. Localization as self-organization of plastic flow

Considering the fundamental possibility of self-organization of a plastically deformable medium, let us begin with the formulation of Haken [2014], according to which: "*A system is called self-organizing if it acquires some spatial, temporal or functional structure without specific external influence*". This statement is rather general in nature and does not contain indications of a self-organization mechanism, which should be considered separately in each specific case.

In addition, arguing the important role of localization, we also give the expression of Knyazevaya and Kurdyumov [1994]: "*The effect of creating structures in an open nonlinear medium is associated with the effect of localization*".

3.1.1. Actual properties of plastically deformable media

Analysis of the applicability of a synergistic approach to the problem of plasticity requires an answer to the question of whether plastically deformable media possess the properties necessary for self-organization, as is provided for in synergy [Glensdorf, Prigogine, 1973; Polak, Mikhailov, 1983; Loskutov, Mikhailov, 1990; Knyazev, Kurdyumov, 1994; Haken, 2014]. In accordance with the point of view of Krinsky and Zhabotinsky [1981], Polak and Mikhailov [1983], Vasiliev, Romanovsky and Yakhno [1987], active, nonlinear, open systems that are far from equilibrium, which exchange energy or matter with the environment, are capable of self-organization. Let us now find out the presence of these qualities in a plastically deformable medium, which we will consider as shown in Fig. 3.1 as a set of elastically deformed volumes (areas of balancing elastic stresses) and elastic stress concentrators existing in it [Barrett, 1948; Roitburd, 1974; Shermergore, 1977; Likhachev, Malinin, 1993].

Hindered flat clusters can serve as a simple model of stress concentrators [Mott, 1951, 1952; Stroh, 1954; Eshelby, 1963; Meyers et al., 2001]. During plastic deformation, the concentrators relax, generating new dislocations and generating slip [Krempl, 2001]. As a result of such an act, the elastic energy and structural defects are redistributed over the volume of the medium with the birth of new concentrators.

The *activity* of a deformable medium having the described structure follows from a comparison of its properties with the definition given by Romanovsky, Stepanova and Chernavsky [1984],

Fig. 3.1. Areas of balancing internal stresses in a deformable medium. Arrows – stress concentrators of different power.

who stated: "*The characteristic features of active kinetic media include the following: a) there is a distributed source of energy or substances rich in energy and negentropy; b) in the medium, one can distinguish an elementary volume of complete mixing, which contains an open point system, far from thermodynamic equilibrium; c) the connection between adjacent elementary volumes is due to the transfer phenomena*".

In the case of a deformable medium, the energy source is the elastic fields of stress concentrators, and the volume of the concentrator can be considered as the volume of complete mixing. As for the relationship between such volumes, its mechanism will be considered later.

The *non-equilibrium* of the deformable medium is caused by two factors. First, it follows from the existence of elastic stress concentrators, due to which the stressed volumes are in states far from the global or local energy minima [Zhirifalko, 1975; Khristian, 1978; Oliferuk, Maj, 2009]. Secondly, the test sample is a system through which mechanical flows occur [Honeycomb, 1972; Zuev et al., 2001; Pelleg, 2013], thermal [Caillard, Martin, 2003] or electromagnetic [Sprecher, Mannan, Conrad, 1986; Zuev, 1990; Molotskii, 2000; Bivin, 2015] energy from external devices. Any such effect on the material or their combination can initiate stress relaxation and a corresponding change in the structure of the deformable medium.

It is shown that the *nonlinearity* of a deformable medium directly follows from the fact that dependence $\sigma(\varepsilon)$ itself, which relates the main variables of the deformation process, the strain ε and stress σ, is non-linear. The nature of the nonlinearity varies continuously along the plastic flow curve and depends on the grade, structure of the material and the mode of deformation. The causes of nonlinearity can be caused by the nonlinearity of the crystal lattice itself [Kittel, 1978; Frenkel, 1972; Newnham, 2005], and the nonlinearity of dislocation ensembles [Aifantis, 1996; Mughrabi, 2001, 2004].

The memory, determined by irreversible processes of structure formation during plastic flow, introduces additional complexity in the description of the properties of a deformable medium [Hill, 1956; Otsuka, Shimizu, 1986; Han Chin-Wu, 2005]. Due to memory, the curve of repeated plastic deformation of the material after unloading never repeats the shape of the original curve σ(ε). Original properties can be restored only by long-term annealing. In addition, elastic and plastic strains coexisting in a deformable medium have a different nature, and the curve σ(ε) describes the behaviour of the machine –sample system, but not of the material, so that its shape is not a property of the material inherent to it.

This combination of properties of a deformable medium explains the complexity of its adequate description. To construct a theory of plasticity, media there are only a few models used in the mechanics of a deformable solid, such as ideally plastic, plastic with hardening, etc. [Hill, 1956; Sedov, 1970; Rabbinov, 1988; Richards, 2001]. The relaxation models of Hooke, Newton, Bingham, Kelvin, Maxwell bodies and their combinations adopted in mechanics and their combinations [Reiner 1962; Sedov, 1970; Novick, Berry, 1975] are not universal and not suitable for constructing an adequate plasticity theory.

Thus, the deformable medium has all the signs of activity and is sufficiently complex for self-organization processes to be possible in it. This opens up perspectives for understanding the nature of the plastic forming process. Microscopic mechanisms of interaction of defects causing the formation of dislocation ensembles have long been developed in the theory of dislocations. So, in the review of Wirtman [1987], long-range (due to elastic fields) and short-range (due to intersection of dislocations) microscopic strengthening mechanisms are considered. However, to explain the large-scale localization of plastic deformation, it is necessary to develop new model concepts.

3.1.2. Hypothesis about the autowave character of localized deformation

A common feature of the self-organization of the environment is the formation of dissipative structures of various forms in it [Nicolis, Prigogine, 1979, 1990]. This process corresponds to the spatial separation of chemical, physical or biological media of different nature. Since the external signs of dissipative structures and patterns of localized plastic flow are very similar, the formation of patterns of localized plastic flow can also be viewed as segregation of a deformable medium with the formation of a heterogeneous spatial-temporal structure, as discussed in Chapter 2.

This allows us to formulate a hypothesis, according to which the *observed spatial-temporal periodic structures – patterns of localized deformation – are dissipative structures*. The concept of dissipative structure, being very general [Glensdorf, Prigogine, 1973; Nicolis, Prigogine, 1979, 1990], requires the introduction of a specific implementation mechanism, in which *autowaves* are most often used [Vasilyev, Romanovsky, Yakhno, 1987]. Therefore, further, the patterns of localized deformation will be considered as modes of localized plastic waves [Zuev, 1994, 1996, 2006, 2007; Zuev, 2001, 2007].

Discussing the meaning of the concept of autowave, we will use its definition given by Krinsky and Zhabotinsky [1981]: "*Self-sustaining waves in active media are usually called autowaves, keeping their characteristics constant due to the energy source distributed in the medium. These characteristics — period, wavelength, amplitude, and form — in the steady state depend only on the local properties of the medium and do not depend on the initial conditions, and far enough from the boundaries of the medium do not depend on the boundary conditions or the linear dimensions of the system. Autowaves generate a macroscopic linear scale due to local interactions, each of which does not have a linear scale.*"

In favor of identifying autowaves and patterns of localized plasticity, there is a similarity, consisting in the fact that:

- patterns of localized plastic flow are fairly stable and exist for a time while the corresponding stage of strain hardening lasts;
- patterns arise with monotonous growth of deformation and do not require for their excitation a special time-varying effect, as is necessary for ordinary waves;

- the patterns have a macroscopic linear scale of l» 10-2 m, arising from the interaction of dislocations having a microscopic scale of the order of the Burgers vector, $b \approx 10^{-10}$ m, with $b \ll \lambda$.

To accept or reject the autowave hypothesis, let us compare the possibilities of an adequate description of the spatial-temporal periodic patterns of plastic flow using one of three possible processes: elastic waves [Lüthi, 2007], plasticity waves [Kola, 1955; Shestopalov, 1958; Davis, 1961; Bell, 1984; Clifton, 1985] and autowaves [Romanovsky, Stepanova, Chernavsky, 1984; Vasiliev, Romanovsky, Yakhno, 1987; Mishchenko et al., 2010]. To begin, let us estimate the speed of propagation of these wave processes, which are:

– for elastic waves $V_t \approx \sqrt{\theta / \rho} \approx (1...3) \times 10^3$ m/s [Lüthi, 2007];

– for plasticity waves V_{pw} $\sqrt{\theta / \rho} \approx (10...10^2)$ m/s [Shestopalov, 1958];

– for patterns of localized plasticity $V_{aw} \approx (10^{-5}...10^{-4})$ m/s [Zuev, Danilov, Barannikova, 2008; Zuev, 2015].

From the comparison of characteristic velocities it follows that $V_{aw} \ll Vpw \ll V_t$. This difference makes us think that each of the processes is generated by its own specific micromechanism. We note that only elastic deformation is directly related to elastic waves, so the choice should be made between *Kolsky plasticity waves and autowaves*.

Based on the work of Madelung [1961], Ango [1967], Jeffreys, Svirles [1969], Krinsky, Zhabotinsky [1981], Osmera [Othmer, 1991], Mishchenko et al. [2010] we compiled Table 3.1, which allows to compare the properties of wave and autowave processes. The data presented in it inclines us to the choice of the two elements of the counter-thesis "waves–autowaves" of the autowave variant of the description of the kinetics of localized plastic flow.

Additional arguments in favor of this choice are:

- statement by Nicolis and Prigogine [1990]: "*Such important and widespread mechanical phenomena as plasticity and fluidity cannot be investigated on a purely mechanical basis! Instead, they should be considered as part of the general problems of nonlinear dynamical systems operating far from equilibrium*",

- opinion of Seeger and Frank [Seeger, Frank, 1987]: "*Increasing the dislocation density and the development of dislocation structures dur-*

ing shear deformation is a process of structure formation. We can assume that the irreversible processes and the collective behaviour of the components of the crystal (atoms, ions, molecules, etc.) associated with such processes play an important role in the deformation'',

- numerous experimental data on the nature of the observed patterns of localized plasticity [Zuev, Danilov, Barannikova, 2008; Zuev, 2006, 2011, 2015; Zuev, 2001, 2007, 2012; Zuev et al., 2004] and their comparison with the structures of autowaves in systems of a different nature [Vasilyev, Romanovsky, Yakhno, 1987; Loskutov, Mikhailov, 1990; Bass, Bakanas, 2000].

It is known that the concepts of specific wave processes arising in active media are not limited only to autowaves of the type being discussed. So Nicolis and Prigogine [1979], as well as Polak and Mikhailov [1983], used the notion of *kinematic waves* (*pseudowaves*), arising from the connection between elements of the active medium, to describe periodic dissipative structures. Depending on the level of this connection, the medium can generate various periodic processes, and pseudowaves arise when the processes are activated in the volumes of the medium located at a macroscopic distance from each other. Apparently, pseudowaves practically do not differ from the discussed autowaves.

3.1.3. Entropy of wave and autowave deformation processes

The choice between waves and autowaves is essentially a choice between systems with energy dissipation (waves) and systems with the formation of dissipative structures (autowaves) [Nicolis, Prigogine, 1979]. The decisive argument here may be the sign of the change in the entropy of the process ΔS [Klimontovich, 1999, 2003]. The case $\Delta S > 0$ corresponds to the dissipative process of energy dissipation, and the case $\Delta S < 0$ corresponds to the process of structure formation (ordering) in the medium.

In the case of deformation processes, the choice can be based on the difference in the type of dependences $V_w(\theta)$ for Kolsky waves ($V_{pw} \sim q^{1/2}$) and autowaves of localization of plastic deformation (1 $V_{aw} \sim \theta^{-1}$). It can be expected that when these types of waves are generated, the nature of the entropy change will be different [Zuev, 2005]. Let the plastic strain rate be $\dot{\varepsilon} \sim V_w$, where V_w is the speed of one of the types of waves. Using the Taylor–Orowan equation (1.1) and assuming that the density of mobile dislocations is $\rho_m \approx$

const, we conclude that $V_{disl} \sim \dot{\varepsilon} \sim V_w$. The rate of thermally activated dislocation movement is determined by the relation (1.2) [Indenbom, Orlov, Estrin, 1978]. Then, taking into account formula (1.2), the velocity of propagation of a deformation wave of any type [Engelke, 1973; Caillard, Martin, 2003]

$$V_w \sim \dot{\varepsilon} \sim V_{disl} \sim \exp\left(\frac{S}{k_B}\right)\exp\left(-\frac{U-\gamma\sigma}{k_B T}\right) \quad (3.1)$$

Most monocrystals and polycrystals studied have similar mechanical characteristics under linear strain hardening. Then, with acceptable accuracy, we can assume that $(U-\gamma\sigma)/k_B \approx$ const. Under this condition, it is obvious that $\ln \dot{\varepsilon} \sim \ln V_w \sim S$. Consequently, the dependences $V_w(\theta)$ for both types of wave processes, shown in Fig. 3.2 in the coordinates V_w–$\ln \theta$ qualitatively correspond to the dependences S–$\ln \theta$ for them.

Data in Fig. 3.2 for the speed of strain localization autowaves at the stages of easy slip of single crystals of Cu, Sn, Fe_I and linear strain hardening of single crystals and polycrystals of metals and alloys are taken from experimental data, and the propagation rates of plasticity waves are calculated as $V_{pw} \approx \sqrt{\theta/\rho}$ using the values of θ calculated by loading curves and density ρ data for the materials studied.

From Fig. 3.2 it follows that with the generation of Kolsky waves, the entropy of the deformed system increases ($\Delta S > 0$), which is typical of dissipative processes. On the contrary, with the generation of autowaves of a localized plastic flow, the entropy decreases ($\Delta S < 0$). This is the final argument in favor of the description of the detected patterns of localized plastic deformation in the language of autowave processes, which serve as the basis for the self-organization processes in the deformable system [Davydov et al., 2004]. At present, the autowave approach to describing the process of plastic flow has been successfully developed by many authors. This refers to theoretical [Zaiser, Aifantis, 2006; Aifantis, 2014] and to the experimental [Ohashi, Kawamukai, Zbib, 2007; Acharia et al, 2008; Fressengeas et al, 2009; Mudrock et al., 2011; Lebyodkin et al., 2012; Roth, Lebedkina, Lebyodkin, 2012; Tretyakov, Wil'deman, 2016] research of this problem.

3.2. Autowave plastic flow equations

The next step, obviously, should be to write the equations of

Table 3.1. Comparison of wave and autowave processes

Waves	Autowaves
Being solutions of	
Hyperbolic differential equations in partial derivatives of the type $\ddot{y} = c^2 y^n$	Parabolic differential equations in partial derivatives of the type $\dot{y} = \zeta(x)+Qy^n$
Describe propagation of	
irreversible perturbation processes in a passive medium	irreversible processes of self-organization in an active medium
Linear scale of the process is determined by	
boundary and initial conditions	microscopic interactions of elements of the medium
Propagation velocity is determined by	
properties of the medium	is independent of properties of medium
Are excited by	
external effect	internal interactions in medium
Typical examples	
elastic waves, electromagnetic waves	switching autowaves, phase autowaves

autowave processes of plastic flow. Since the general theory of autowave processes has been developed in sufficient detail [Vasi Lion, Romanovsky, Yakhno, 1987; Mishchenko et al., 2010], the mathematical form of the corresponding equations is, in principle, known and the main difficulty is their adaptation to the processes of plastic deformation.

3.2.1. On the structure of autowave equations

Parabolic differential equations of the second order in partial derivatives of the type $\dot{y} = \zeta(x)Qy^n$ (see Table 3.1), describing autowave processes, are formally obtained by adding a nonlinear function $\zeta(x)$ to the right side of the equation $\dot{y} = Dy^n$, similar to Fourier equation for thermal conductivity $\dot{T} = \kappa T^*$ [Madelung, 1961; Carsloe, Eger, 1964] or the 2nd Fick equation for diffusion $C = DC^n$ [Manning, 1971; Merer, 2011]. Here, κ is the coefficient of thermal diffusivity, C is the concentration, and D is the diffusion coefficient. This was first done by Kolmogorov, Petrovsky, and Piskunov [1937], who wrote an equation to describe the kinetics of processes in nonlinear media as applied to biological problems (authors' notation)

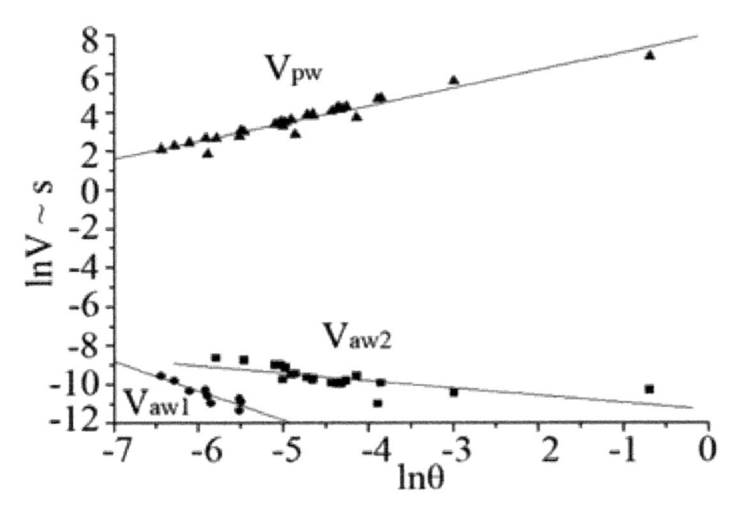

Fig. 3.2. Entropy change during the formation of plasticity waves and autowaves of plastic flow localization at the stage of easy slip V_{aw1} and at the stage of linear strain hardening V_{aw2}.

$$\frac{\partial v}{\partial t} = k\left(\frac{\partial^2 v}{\partial x^2} + \frac{\partial^2 v}{\partial y^2}\right) + F(v). \quad (3.2)$$

Here x and y are the coordinates of a point on the plane; $k > 0$ is the diffusion coefficient; v is the density of matter at point (x, y) at time t, and $F(v)$ is point kinetics, a non-linear N-shaped function that has the meaning of the local rate of change of v in the total mixing volume. Reaction-diffusion equations of type (3.2) are basic for synergetics [Klimontovich, 2002].

The introduction of the nonlinear function $F(v)$ into equation (3.2) is not a purely formal operation: it is due to the appearance of this function that conditions are provided for excitation in a homogeneous system of autowaves and for the system to acquire macroscopic scales [Mishchenko et al., 2010]. The explicit form of this function can be obtained either from physical considerations using a specific mechanism of the phenomenon, or in the form of a polynomial with a cubic nonlinearity [Zykov, 1984]. The second option has significant mathematical advantages, but is inferior to the first one due to the difficulty of physical interpretation of the function recorded in this way.

The presence of the first time derivative in equation (3.2) underlines the irreversible nature of the processes described by it [Chester, 1966; Ziegler, 1966]. Instead of the diffusion coefficient k, when writing such equations, the value of $L^2/\vartheta = k$ is often used, which simultaneously includes the characteristic scale L and the relaxation time ϑ of the corresponding factor.

Equation (3.2) and the basic techniques of its solution are developed in the book by Mishchenko et al. [2010] and are widely used in analyzing the processes of the birth of structures in chemical and biophysical objects. They also found a number of other important and interesting applications. So, for example, in the works of Samara et al. [1976; 1977] similar equations are involved to describe various nontrivial regimes arising during combustion. Nekorkin and Kazantsev [Nekorkin, Kazantsev, 2002] studied the pattern of propagation of autowaves in a three-component reaction mixture. Zaikin and Morozova, 1978], Elenin et al. [1983], Davydov, Zykov, Mikhailov [1991], Ataullakhanov et al. [2002, 2007], Davydov et al, [2004], Tsyganov et al. [2007], and Zemskov and Loskutov [2008] proposed solutions for autowave equations for active media of different nature and considered the development of autowave processes in the form of moving fronts.

In the physics of defects, one of the first attempts to use equation (3.2) belongs to Dont [1963], who explained the role of redistributing kinks on a dislocation line in amplitude-dependent internal friction of metals. At present, such ideas are beginning to be widely used to describe the self-organization of dislocation ensembles during plastic deformation. For example, Khannanov [1992] and Khannanov and Nikanorov [2007] obtained a correct estimate of the propagation speed of localized plasticity for the case of the front of the Chernov–Lüders band in the framework of a simple dislocation model.

3.2.2. Equations of autowaves of localized plastic flow

In the general case, the existing models of autowave processes take into account the competition between autocatalytic (activator) and damping (inhibitor) control factors [Polak, Mikhailov, 1983; Vasiliev, Romanovsky, Yakhno, 1987]. This means that an adequate description of complex systems requires at least two equations. In problems of chemical kinetics, for example, one of the equations describes the concentration of the activator, and the other - the inhibitor. In systems of a different nature such substances may not be. Therefore, the choice of specific factors requires at least a qualitative understanding of the nature of the phenomena in the system under study.

When solving the problem of localized plastic flow of solids in [Zuev et al., 1998; Zuev, Danilov, 1999; Zuev, 2012] as an activator, it was proposed to consider deformations as an inhibitor – elastic stresses. The main argument in favor of this choice is as follows. As it is known [Hill, 1956, Rabotnov, 1988], the stress tensor in describing the deformation of the spinal environment is usually divided into deviator and spherical parts. The tensor-deviator is responsible for the actual plastic shaping, and the spherical tensor creates hydrostatic compression or tension and prevents plastic deformation in accordance with the law of elasticity of volume strain, [Rabotnov, 1988].

In addition, the sample is adiabatically cooled during elastic tension, and when dislocation deformation, local heat generation occurs on shear planes (see, for example, [Klyavin, 1987]). The arising two temperature effects accompanying the plastic flow are opposite in their influence on the process.

The choice of plastic deformation and elastic stress as control parameters is obviously convenient because it is possible to experimentally determine the quantities σ and ε directly from the diagram $\sigma(\varepsilon)$. Finally, such a choice allows one to take into account the spatial separation of elastically stressed zones and plastic shear zones, which determines the multi-scale plastic deformation [Malygin, 1991, 1995; Sarafanov, 2001, 2008; Zbib, de la Rubia, 2002; Aifantis, 1992, 2001].

The mechanisms of the action of factors controlling the development of plastic flow can be as described as follows. The autocatalytic factor (deformation) acts in such a way that each completed shift initiates a similar process required for accommodation in the adjacent volume, so that the effective radius of this factor has a shear zone size order l, and its propagation velocity is commensurate with the dislocation velocity V_{disl}. On the other hand, during each act of shear, the released elastic energy is redistributed throughout the volume, causing a relative increase in stress concentration, which complicates plastic deformation and is equivalent to the action of a damping factor. The radius of action of the latter is of the order of the sample size $L \gg l$, and the propagation velocity is the velocity of elastic waves $V_t >> V_{disl}$. It is essential that exactly this ratio between the radii of action and the speeds of propagation of control factors is necessary for the generation of autowaves [Murray, 1983; Vasiliev, Romanovsky, Yakhno, 1987].

In many works in the description of the dislocation mechanisms of plastic deformation [Gilman, 1965; Khannanov, 1992; Aifantis, 1995; Malygin, 1999] the mobile and stationary dislocation densities are often used as the activator and inhibitor. In this case, it is difficult to accurately estimate the behaviour of these quantities in the course of plastic flow [Gilman, 1965], so that solving the obtained equations requires the use of numerous and not always justified assumptions.

Given such a choice of controlling factors as applied to deformation processes, the equations for deformation (activator) and stresses (inhibitor) can be written by analogy with equation (3.2) (actually postulated at this stage) as a system

$$\dot{\varepsilon} = f(\varepsilon) + D_{\varepsilon\varepsilon}\varepsilon'', \tag{3.3}$$

$$\dot{\sigma} = g(\sigma) + D_{\sigma\sigma}\sigma'', \tag{3.4}$$

which describe the rates of change of strains and stresses, and contain nonlinear functions (point kinetics) $f(\varepsilon)$ and $g(\sigma)$, which are usually oscillatory in nature. The meaning of using double subscripts for the coefficients $D_{\varepsilon\varepsilon}$ and $D_{\sigma\sigma}$ in equations (3.3) and (3.4) will be explained in Chapter 4.

Given the importance of equations (3.3) and (3.4), it is necessary to consider the possibility of their derivation from the basic equations of mechanics. Equation (3.3) can be obtained taking into account the condition formulated by Sedov [1975], according to which: "*Functions entering the law of motion of a continuum are continuous and have continuous partial derivatives with respect to all their arguments*." We write on this basis the continuity condition of plastic flow in the form

$$\dot{\varepsilon} = \nabla \cdot (D_{\varepsilon\varepsilon}\nabla\varepsilon), \tag{3.5}$$

where is the strain flow in the deformation gradient field. If $D_{\varepsilon\varepsilon}(x)$, then

$$\dot{\varepsilon} = \varepsilon' \cdot D'_{\varepsilon\varepsilon} + D_{\varepsilon\varepsilon}\varepsilon'' = f(\varepsilon,\sigma) + D_{\varepsilon\varepsilon}\varepsilon'' \tag{3.6}$$

where $f(\varepsilon,\sigma) = \varepsilon' \cdot D'_{\varepsilon\varepsilon}$ is a non-linear function of strain and stress, and the coefficient $D_{\varepsilon\varepsilon}$ has the dimension $L^2 \cdot T^{-1}$.

In turn, equation (3.4) for stresses follows from the Euler equation for a viscous fluid flow, written in the form

$$\frac{\partial}{\partial t}\rho v_i = -\frac{\partial \Pi_{ik}}{\partial x_k} \tag{3.7}$$

[Landau Lifshits, 2001]. In a viscous medium,

$$\Pi_{ik} = p\delta_{ik} + \rho v_i v_k - \sigma_{vis} = \sigma_{ik} - \rho v_i v_k$$

is the pulse flux density tensor, δ_{ik} is the unit tensor, p is the pressure, and v_i and v_k are the components of the flow velocity. The stress tensor $\sigma_{ik} = -p\delta_{ik} + \sigma_{vis}$ is the sum of elastic $\sigma_{el} = -p\delta_{ik}$ and viscous stresses σ_{vis}, that is, $\sigma = \sigma_{el} + \sigma_{vis}$ and $\dot{\sigma} = \dot{\sigma}_{el} + \dot{\sigma}_{vis}$. The relaxation rate of elastic stresses [Dotsenko, Landau, Pustovalov, 1987] $\dot{\sigma}_{el} \equiv g(\sigma,\varepsilon) = -\frac{M\rho_m b^2}{B}\sigma = -M\rho_m bV_{disl} \sim V_{disl}$, where M is the elastic modulus of the sample–testing machine system, $B \approx 10^{-5}$–10^{-4} Pa·sa is the coefficient of quasi-viscous deceleration of dislocations, and $V_{disl} = (b/B)\cdot\sigma$ is the rate of their viscous motion [Alshitz, Indenbom, 1975a, b].

Viscous stresses σ_{vis} are linearly related to changes in the propagation velocity of elastic waves in the medium during deformation $V_t \approx V_{t0} + \beta\sigma$ [Muravyov, Zuev, Komarov, 1996], where V_{t0} is the propagation velocity of transverse elastic waves in the absence of stresses, and β = const. If we assume that $\sigma_{vis} = \eta\nabla V_t$, where η is the dynamic viscosity of the medium, then $\partial\sigma_{vis}/\partial t = V_t\nabla\cdot(\eta\nabla V_t) = \eta V_t\,\partial^2 V_t/\partial x^2$. Then the relaxation rate of viscous stresses is $\partial\sigma_{vis}/\partial t = \eta V_t\,\partial^2 V_t/\partial x^2 = \eta\beta v_t\partial^2\sigma/\partial x^2$

$$\partial\sigma/\partial t = g(\varepsilon,\sigma) + D_{\sigma\sigma}\,\partial^2\sigma/\partial x^2 \tag{3.8}$$

where $D_{\sigma\sigma} = \eta\beta V_t$ is the transfer coefficient with dimension $L^2 \cdot T^{-1}$.

On the right side of equation (3.8) $\dot{\sigma} = g(\sigma)$ and $\dot{\sigma}_{vis} = D_{\sigma\sigma}\,\partial^2\sigma/\partial x^2$ are the relaxation rates of elastic and viscous stresses, respectively. But, if the nonlinear function $g(\sigma)$ takes into account the redistribution of elastic stresses between adjacent microvolumes on the front of an existing plasticity source, then the member $D_{\sigma\sigma}\,\partial^2\sigma/\partial x^2 \equiv \eta\zeta V_t\,\partial^2\sigma/\partial x^2$ is responsible for the processes of redistribution stresses in the sample on a macroscopic scale.

3.2.3. Analysis of autowave equations

In the general case, the flow of plastic deformation includes hydrodynamic and diffusion components. The first one is described by nonlinear functions $f(\varepsilon,\sigma) \sim V_{disl}$ and $g(\sigma,\varepsilon) \sim V_{disl}$ in equations (3.3) and (3.4), associated with the continuous propagation of deformation along the sample due to the successive relaxation of local stress concentrators. It is realized in the form of continuous motion along the sample of the deformation front with successive activation of local stress concentrators on it. The second component is determined by the diffusion terms $D_{\varepsilon\varepsilon}\partial^2\sigma/\partial x^2$ and $D_{\varepsilon\varepsilon}\partial^2\sigma/\partial x^2$. It is associated with dislocation shifts during relaxation of one of the stress concentrators formed at the previous stages of the process. This process ensures the formation of a deformation zone at a macroscopic distance ~λ from the already existing plasticity front. Due to this, there is a kind of throwing deformation at a macroscopic distance, which is the source of the macroscopic scale of plastic flow ~λ.

The meaning of equation (3.6) for the rate of plastic deformation can be clarified by establishing its relationship with the Taylor-Orowan equation of dislocation kinetics (1.1). To do this, we consider the function $f(\varepsilon) = \varepsilon' D'_{\varepsilon\varepsilon}$ in the right-hand side of equation (3.7) with a low density of mobile dislocations. If the latter are uniformly distributed over the average distance l, then bl is the shear deformation when one dislocation is shifted by l, and l^{-2} and ρ_m is the density of mobile dislocations. Then $D_{\sigma\sigma} \cdot \partial^2\sigma/\partial x^2 \equiv \eta\varsigma V_t \cdot \partial^2\sigma/\partial x^2$ and using the diffusion representation $D_{\varepsilon\varepsilon} \approx L_{disl} V_{disl}$, where $L_{disl} \approx \alpha x$ is the length of the path of the dislocations, and V_{disl} = const is their speed, we get:

$$\dot{\varepsilon} = \alpha b \rho_m V_{disl} + D_{\varepsilon\varepsilon}\varepsilon''. \tag{3.9}$$

The dimensionless coefficient $\alpha = L'_{disl} \approx$ const, so that the first term in the right-hand side coincides with the Taylor–Orowan equation (1.1), which can now be considered as a special case of equations (3.6) and (3.9), suitable for describing 'point kinetics' of a deformed system – a relaxation act, usually proceeding in the form of a nonlinear process of jump-like deformation [Pustovalov, 2008]. Comparing equations (1.1) and (3.9), it follows that equation (1.1) describes only the hydrodynamic component of the deformation flux, ignoring the possibility of the formation of defects throughout the volume due to the contribution of the diffusion-like terms $D_{\varepsilon\varepsilon}\varepsilon''$ and $D_{\varepsilon\varepsilon}\sigma''$. This possibility is taken into account in the framework of the more general equation (3.9), which in the right-hand part contains two components of the flow of plastic flow of different nature: hydrodynamic and diffusion.

In the diffusion component of the deformation flux $D_{\varepsilon\varepsilon}\varepsilon''$, the value of $D_{\varepsilon\varepsilon} \approx L_{disl} \cdot V_{disl}$ at a constant shear stress remains almost unchanged. However, the magnitude ε'' can be large due to the non-uniformity of the distribution of defects. From the theory of parabolic differential equations [Ango, 1967; Jeffries, Swirles, 1969], to which equation (3.6) relates, it is known that the transmission rate of interaction between elements of the medium in such systems is in $V^{(in)} \to \infty$. Such a situation is physically unrealizable and should be replaced by the condition $V^{(\text{in})} \approx V_t$, limiting the maximum transmission rate of signals in a continuous medium to the speed of sound V_t. Thus, in the limit, the deformation processes should be continuing at the speed of sound (the speed of propagation of elastic waves). In this case, of course, we do not deal with dislocation distances over distances commensurate with the sample size. Without discussing the nature of the generation of shifts at a macroscopic distance from the active deformation zones, one can think that it is acoustic phenomena that can play an important role in this process.

With the help of equations (3.3) and (3.4), we now analyze the possible modes of plastic deformation by linearizing and constructing the phase portrait of the system. Specific types of functions $f(\varepsilon, \sigma)$ and $g(\varepsilon,\sigma)$ for N-shaped point kinetics of relaxation deformation processes were proposed earlier [Zuev, Danilov, 1999] in the form

$$f(\varepsilon,\sigma) = -\frac{\varepsilon}{\theta_\varepsilon} + \frac{\sigma}{\eta}, \tag{3.10}$$

$$g(\varepsilon,\sigma) = -\frac{(\sigma - \sigma_y) - \sigma^*}{\theta_\sigma} + \frac{\sigma\varepsilon}{\theta_\varepsilon}. \tag{3.11}$$

Here η is the dynamic viscosity of the deformable medium; $\theta_\sigma \ll \theta_\varepsilon$ are the relaxation times of elastic stresses and strains, respectively; σ^* is the stress of the end of relaxation, σ_y is the yield strength. Using equations (3.11) and (3.12), the nonlinear properties of the deformable medium are taken into account. To analyze the deformation processes, we construct the 0-isoclines of equations (3.10) and (3.11), equating their right-hand sides to zero. In this case

$$\sigma = \frac{\eta}{\theta_\varepsilon}\varepsilon = G\varepsilon, \tag{3.12}$$

$$\sigma = \frac{\sigma^* + \sigma_y}{1 - (\theta_\sigma/\theta_\varepsilon)\varepsilon}, \tag{3.13}$$

as shown in Fig. 3.3.

A qualitative analysis of equations (3.12) and (3.13) consists in finding the intersection points of the 0-isoclines. It is clear that equation (3.12) is simply Hooke's law. The N-shaped form and position of the function (3.13) is determined by the ratio of the constants included in it and the level of the current strain ε, and $\sigma^* = \sigma^*(\varepsilon)$. In the region of low stresses (0-isocline 1 in Fig. 3.3), after reaching the intersection point Ω, any small deviation from

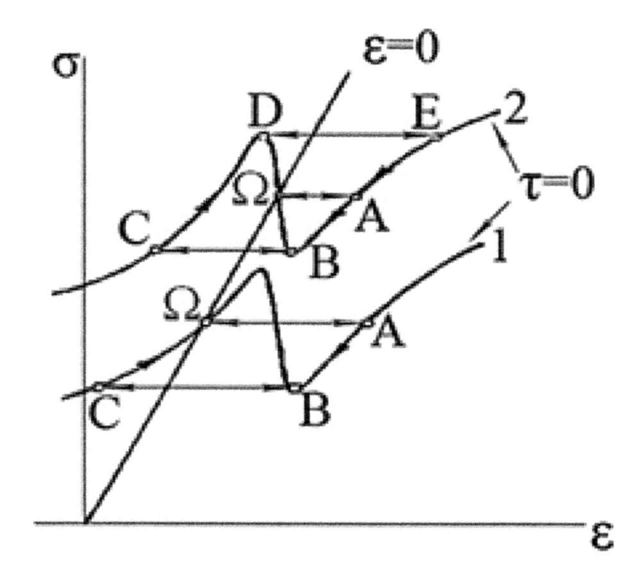

Fig. 3.3. 0-isoclines of equations (3.13) and (3.14).

equilibrium leads to an abrupt $\Omega \rightarrow A$ transition to a stable isocline branch $\dot{\sigma} = 0$. At low stresses, the imaging point performs an $A \rightarrow B \rightarrow C \rightarrow \Omega$ cycle, and the system comes to equilibrium again. This case corresponds to the propagation of a single pulse in the system, which, obviously, can be identified with the front of the Chernov-Lüders band.

At a higher stress level (0-isocline 2 in Fig. 3.3) after the breakdown $\Omega \rightarrow A$, the imaging point does not return to the equilibrium position, but moves along a closed cycle A→ B→ C→ D→E→B. Repetition of the cycles in this case indicates a coordinated movement of regions forming the phase autowave of strain localization. The considered scheme not only explains the existence of two different autowave modes, but also provides for the possibility of rebuilding the autowave modes with an increase in the flow voltage, which is observed experimentally.

The development of autowave structures, due to the presence of strong multi-scale correlations in an ensemble of elementary carriers of plastic deformation (dislocations), was considered by Hon et al. [2008]. It was shown that to take into account the contribution of large-scale changes in the internal structure to the total deformation, additional variables are required associated with large-scale correlations in the ensemble of defects. As such, the order parameters *p, q* are introduced, representing the volume fractions of the deformed and non-deformed structure characteristic of the hierarchical levels under consideration. This technique is also standard for synergetics [Haken, 2014]. The order parameters are determined by solutions of a system of nonlinear equations of the reaction-diffusion type, similar in form to equations (3.3) and (3.4)

$$t_p \partial p/\partial t = a_p p + b_p p^2 - p^3 + cpq + l_p^2 \partial^2 p/\partial x^2, \quad (3.14)$$

$$t_q \partial q/\partial t = a_q q + b_q q^2 - q^3 + cpq + l_q^2 \partial^2 q/\partial x^2, \quad (3.15)$$

where the coefficients a_a and a_q be both positive and negative, and b_p >, 0 b_q> 0, $c > 0$, $d > 0$. The correlation lengths l_p, l_q characterize the size of the regions of localized deformation, and the relaxation times l_p, l_q determine the stress relaxation rate.

The analysis showed that two types of time-dependent solutions describing the transition of the system from an unstable to a stable state are possible. The first solution is an autowave in the form of a running front, in front of which the medium is in the p_0, q_0 state, and behind it is in the p_h, q_h state, as shown in Fig. 3.4 a. The speed of the front is determined by the values $v_p = lp/tp$, $v_q = l_q/t_q$. In the one-dimensional case, this corresponds to the movement of the deformation fronts, as is observed, for example, with the growth of the Chernov–Lüders bands. At the periphery of the strip, the solutions become stationary and homogeneous.

The second type of solution describes localized nonequilibrium regions, which can be static, running, or pulsating. A homogeneous solution is stable with respect to small perturbations. Autowaves are driven by inhomogeneous perturbations of finite amplitude. In this case, a system of moving or stationary bands of localized deformation should occur, as is observed at the stages of linear and parabolic deformation hardening (Fig. 3.4, b).

Presented in Fig. 3.4 solutions of equations (3.14) and (3.15) have a simple interpretation. The localized plastic flow is realized where the maxima of both order parameters coincide, that is, the growth of one of them initiates the growth of the other.

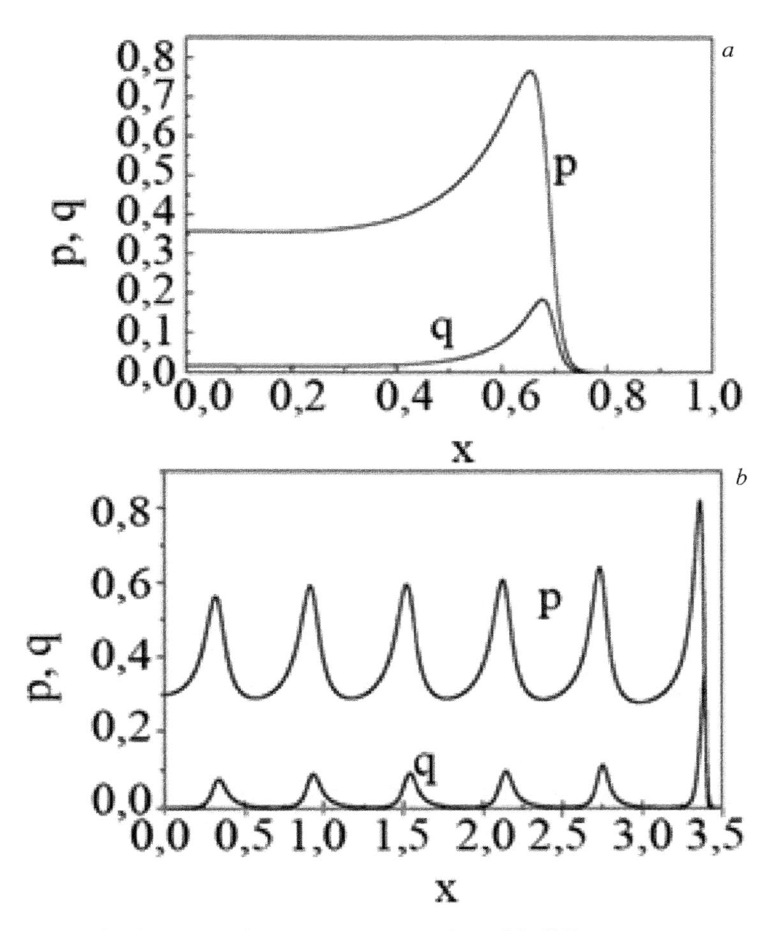

Fig. 3.4. Behaviour of order parameters p and q with different autowave generation modes: switching autowave (*a*); phase autowave (*b*).

3.3. Generation of autowave plastic deformation modes

The occurrence of localized plastic flow autowaves is a common effect for all cases of deformation. It can be assumed that their birth is associated with the action of stress concentrators, which play an important role in the development of the flow process.

3.3.1. Autowave generation by stress concentrators

To test this assumption, special experiments were carried out [Danilov, Narimanova, Zuev, 2000] in order to determine the causal connection between localized plastic flow autowaves and macroscopic concentrators, which were previously grown fatigue cracks.

The origin of plastic flow near the end of the crack has been studied by many researchers [Meyers e tal., 2001; McDonald, Efstathiou, Kurath, 2009]. This problem was considered in detail by Higashida et al. [Higashida et al., 1997, 2004, 2008], who observed the work of Frank–Read sources in these zones in single crystals of Si and MgO using electron microscopy.

In our experiments, a macroscopic concentrator was modelled by a fatigue crack grown in a ductile steel. The task of the study was to *in situ* analyze the shape and size of the plasticity zone before a crack started to develop during ductile fracture and to establish the regularities of the dynamics of plastic flow of material in it. Samples for tests measuring 90×20×10 mm with were cut from a sheet of low carbon steel (0.08 wt.% C) with a thickness of 10 mm. They had a V-shaped notch in the middle of a depth of 8 mm with a radius of curvature at the apex of 0.2 mm. At the top of the notch, a fatigue crack ~2 mm long was grown by cyclic bending. The samples prepared in this way were loaded according to the three-point bending scheme.

When bending in the sample, successively every 100 μm displacements of the movable support by the method of double-exposure speckle photograph on the ALMEC setup, the fields of displacement vectors of the points of the studied surface were recorded. The fields of displacement vectors were recorded step by step from the beginning of the plastic stage of the sample bending to fracture. The accuracy of the measurement of the displacement vector in this case was ~1 μm. The decryption program made it possible to obtain data files, each of which contained a set of displacement vectors ***u*** and the corresponding coordinates. The vector was given by its modulus and angle α relative to the x-axis of the sample. The position of the crack tip $O(x_0, y_0)$ was fixed in the same coordinate system.

In accordance with the concepts of linear mechanics of fracture [Irwin, Paris, 1976; Syratori, Miyoshi, Matsushita, 1986] the field of displacement vectors near the crack end in the polar coordinates r and φ, chosen due to the symmetry of the problem, has the form

$$u(r,\phi)=\frac{K_1\sqrt{r}}{Gk}[(2k+1)^2\cos^2\frac{\phi}{2}+2(4k^2-1)\cos\frac{\phi}{2}\cos\frac{3\phi}{2}+(2k-1)^2\sin^2\frac{\phi}{2}+$$

$$+2(2k-1)^2 \sin\frac{\phi}{2}\sin\frac{3\phi}{2}+(2k-1)^2]^{1/2}. \quad (3.16)$$

Here r is the distance measured from the tip of the crack O; φ is the angle between the direction of the ray r and the direction of the continuation of the crack y; G is the shear modulus; K_1 is the stress intensity factor; k is a constant depending on the Poisson's ratio of the material and the type of stress state. Equation (3.16) shows that in the elastic region, $\sim r^{1/2}$. Obviously, this pattern should be violated at the border of the elastic and plastically deformed zones, which allows to identify the position of such a border using a special search procedure.

We measured 100 values of displacement vectors $\boldsymbol{u}$ along each of the 60 rays r_i with a step of 3° in angle φ. Further, the distances from the vertex r^* were determined, at which the dependence $u(r^{1/2})$ becomes, according to (3.16), linear. As a linearity criterion, we used the correlation coefficient of the experimental dependences $u(r^{1/2})$ interpolated by the least squares method. The current value of u was consistently eliminated as it moved away from the crack tip, until the maximum value of the correlation coefficient was reached. This procedure was carried out for each of the 60 rays r_i. As a result, a set of boundary values r^* was obtained, such that in the region of $r>r^*$, the regularities of linear fracture mechanics are fulfilled, and the geometrical location of the points r^* determines the boundary between the elasticity and plasticity zones in the sample.

The size and shape of the plastic zone gradually change as the strain increases [Reiser, 1970; Botvina, 2008; Pestrikov, Morozov, 2012]. At first, the plastic zone develops mainly along the direction of the initial crack y. In addition, near the yield point, it is abnormally elongated in directions φ = 20–25°. The observed distribution of the plasticity zone boundary is close in nature to the known fronts of the Chernov-Lüders band, which develop in materials with a sharp yield point [Mac Lin, 1965; Honeycomb, 1972; Hallai, Kyriakides, 2013]. The fracture of the boundaries of plastic zones is explained by the spatial heterogeneity of the plastic flow. As the results show, plastic deformation is also heterogeneous in time. Local bands of macroscopic shear can be inhibited and disappear in some directions, but they can arise and develop in others. Therefore, an abrupt displacement of portions of the plasticity boundary is observed before the crack. The heterogeneity of the propagation front of plastic deformation, which differs from the flat Lüders front under tension, is a consequence of a more complex stress state at the crack tip.

Thus, the development of the plastic zone at the tip of a normal fracture occurs non-uniformly both in the sample space and in the load time. In this case, in the early stages of loading, the spatial inhomogeneity is more pronounced. The reason for this behaviour is obviously the already mentioned fact of non-uniform plastic strain distribution in the area of the stress concentrator.

Important are data on the behaviour of the material in the plasticity zone near the concentrator, where the ratios of linear fracture mechanics are not fulfilled and the material is in a plastically deformed state. The behaviour of the material in this zone during loading is described by the evolution of the components of the plastic strain tensor recorded in polar coordinates for the plane case

$$\varepsilon_{ij} = \begin{vmatrix} \varepsilon_{rr} \varepsilon_{r\phi} \\ \varepsilon_{\phi r} \varepsilon_{\phi\phi} \end{vmatrix}. \tag{3.17}$$

The components of the tensor (3.17) can be found from the experimental data on the array of the vector field $\boldsymbol{u}(r, \varphi)$ by numerical differentiation of the components u_r and u_φ of the vector $\boldsymbol{u}$ by the coordinates r and φ for each sample point, where its values for different loading moments are determined. They have the form $\varepsilon_{rr} = \partial u_r / \partial r$, $\varepsilon_{\phi\phi} = 1/r\left(\partial u_\phi / \partial \phi\right) + u_r / r$ and $\varepsilon_{r\phi} = \varepsilon_{\phi r} = 1/2\left[1/r\left(\partial u_r / \partial \phi\right) + \partial u_\phi / \partial r - u_\phi / r\right]$

For the analysis, the distribution of the component ε_{rr} in the x_0 plane for different stages of deformation was used as the most informative characteristic, an example of which is shown in Fig. 3.5. It can be seen that in the plasticity zone in front of the crack (stress concentrator) an ordered spatially periodic system of localized deformation regions with a distance between the maxima of ~2...3 mm is born. Over time, as the strain increases, the amplitude of the inhomogeneous local deformations increases, and the number of such regions grows.

Such a picture can be considered as the initial stage of the development of the autowave of localization of the plastic flow. The shape of the generated autowaves changes with increasing strain, which is apparently due to the heterogeneity of the displacement field created by the crack-type concentrator.

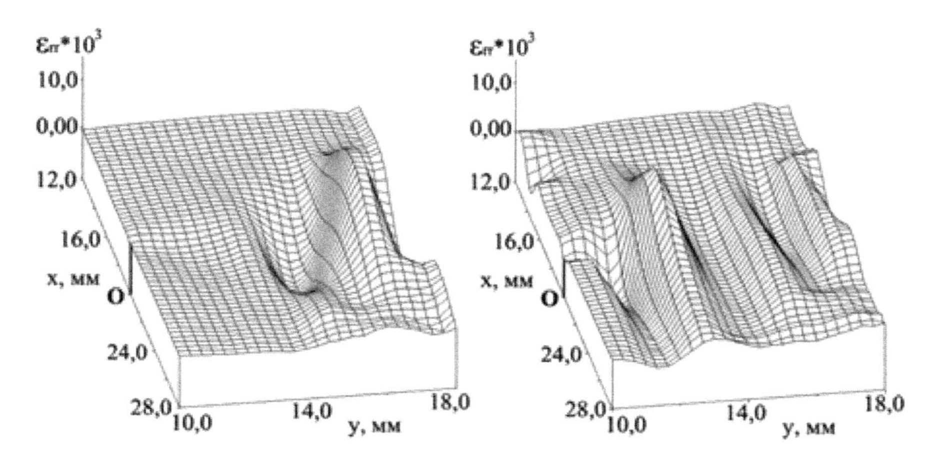

Fig. 3.5. The origin of autowaves of localized deformation at the stress concentrator O. The time between frames 12 s.

3.3.2. Autowave modes of localization of plastic flow

Experimental data show that the patterns of localized plastic flow depend on the strain hardening acting at the appropriate stage of the law, and, as follows from Table 2.4, there are only four types of deformation patterns. Taking into account the ideas about the nature of the stage of strain hardening [Seeger, 1960; Wirtman, Wirtman, 1987; Kuhlman-Wilsdorf, 2002], the characteristics of the behaviour of a deformable material during the formation of corresponding patterns at different stages of hardening, the patterns of localized plasticity can, as mentioned above, be uniquely identified with dissipative structures that were introduced by Nicolis and Prigogine [1979; 1990] to describe the ordering processes in open nonequilibrium systems, or with different modes of *autowave processes* [Vasil'ev, Romanovsky, Yakhno, 1987; Loskutov, Mikhailov, 1990].

At the yield stage stage, when σ = const, the localization pattern has the shape of a moving front of the Chernov–Lüders band separating the deformed and non-deformed parts of the sample [Zuev, 2017]. At the front, elements of the medium irreversibly change from a metastable (elastic) to a stable (plastically deformed) state, and the reverse transition is impossible due to the memory of the medium. When meeting, the fronts annihilate, which makes it possible to interpret them as autowaves, and not as solitons [Kerner, Osipov, 1989; Zakharov, Kuznetsov, 2012]. Therefore, we can assume that the localized deformation at the flow area corresponds to a switching autowave [Loskutov, Mikhailov, 1990], which occurs in an active medium consisting of bistable elements. A similar approach to describing the strain at the yield point was suggested by Gehner [Hähner, 1994].

At the stages of *light slip and linear strain hardening*, for which $\sigma \sim \varepsilon$ and θ = const, the observed pattern is a combination of equally located at distance λ from each other, of localized plasticity. All of them consistently move along the sample with the same speed V_{aw}. At this stage, the condition of constancy of the phase $\omega t - kx$ const, where the frequency $\omega = 2\pi/T$ and the wave number $k = 2\pi/\lambda$ 1 2 are obtained from X–t diagrams of the corresponding processes, is satisfied. The observed picture corresponds to the *phase autowave* [Loskutov, Mikhailov, 1990]. MacDonald, Efstasthiou, and Kurath, 2009, also observed wave phenomena during deformation of copper single crystals.

The mechanism of generation of phase autowaves assumes that the medium consists of self-oscillating elements, in which you can consider, for example, parts of the boundaries between differently stressed or deformed areas [Bell, 1987; Nemes, Eftis, 1992]. The result of numerical simulation of the generation process is presented in Fig. 3.6.

The observed pattern was obtained for the section of the boundary between the bistable and self-oscillating media [Vasil'ev, Romanovsky, Yakhno, 1987]. In the case of plastic deformation, the deformation at the yield point (the front of the Chernov–Lüders band) corresponds to a bistable medium, and the self-oscillation stage corresponds to the stage of linear strain hardening (phase autowave). Thus, the situation shown in Fig. 3.6, can be realized at the transition from the yield point to the stage of linear strain hardening.

The pattern of localized plasticity observed at the stage of *parabolic strain hardening* at $\sigma \sim \varepsilon^{1/2}$ and $\theta \sim \varepsilon^{1/2}$ can be interpreted as a stationary dissipative structure [Nicolis, Prigogine, 1979]. The regions of localized plastic flow formed at this stage are immobile. Such a pattern is similar to a phase autowave, but for it is $V_{aw} = 0$. Distances between plasticity centres are also constant and are $\sim 10^{-2}$ m. A stationary dissipative structure is possible with full synchronization of self-oscillating elements and the establishment of a single phase of self-oscillations throughout the entire deformable sample throughout throughout the stage of parabolic strain hardening.

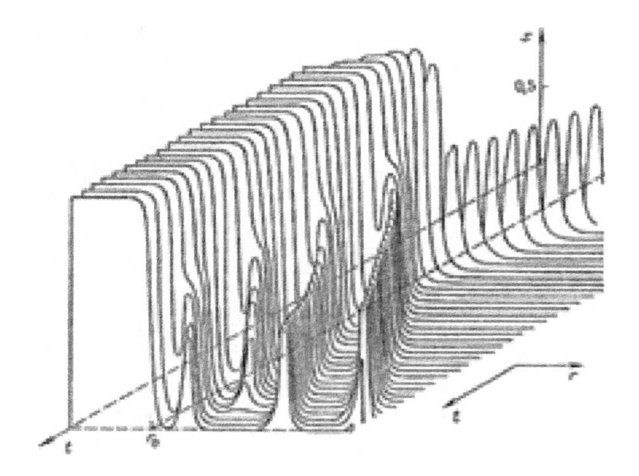

Fig. 3.6. Generation of a phase autowave by the boundary between two active media [Vasilyev, Romanovsky, Yakhno, 1987]

At the *pre-fracture* stage at $\sigma \sim \varepsilon^m$ nd m <½, when a macroscopic neck begins to form in the sample, the localization autowave pattern demonstrates the collapse of the autowave process [Kadomtsev, 1997; Zakharov, Kuznetsov, 2012]. At this stage, the regions of localized plasticity move in a self-consistent manner, so that beams of straight lines on the X–t are regularly formed, as was shown in Fig. 2.18.

The listed autowave modes were observed in all studied metals and alloys. The corresponding list of materials is given in Table 3.2. An analysis of its data shows that autowave patterns of localized plastic deformation represent a phenomenon that is sufficiently common for plastic flow processes.

3.3.3. Deformation as an evolution of autowave structure

The consequence of the point of view, according to which patterns of localization of the plastic flow are defined autowave modes, is the conclusion that plastic deformation is a process of evolution of autowaves of localization of the plastic flow.

As shown in Fig. 3.7, certain autowave processes correspond to the stages of strain hardening, which occur during plastic flow in the following order: elastic wave → switching autowave → phase autowave → stationary dissipative structure → collapse of autowave. Of course, it is usually possible to observe all four autowave modes during deformatioYn of the material since all four stages

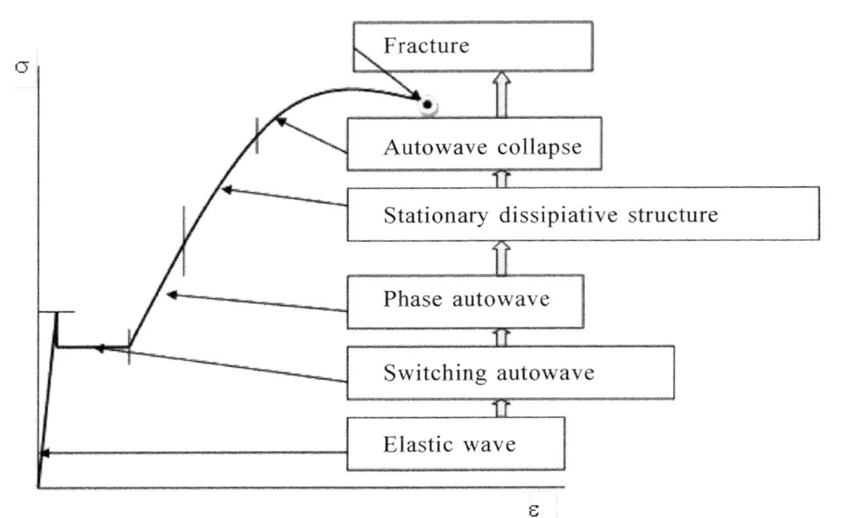

Fig. 3.7. Representation of a multi-stage process of plastic flow as a regular evolution of autowaves of localized plastic deformation.

of strain hardening do not coexist in one material [Zuev, Danilov, Barannikova, 2008].

The transition to autowave representations makes it possible to summarize the Compliance rule obtained in Chapter 2 on the basis of an analysis of the forms of patterns of localized plasticity. In the light of what has been said about the autowave nature of the localization of plastic flow, it is complemented by the notions of autowave deformation modes, as is done in Table 3.3.

It is interesting to compare the revealed patterns of the change of autowave modes during plastic deformation with the autowave processes developing in systems of a different nature. It is known that for modeling each type of autowave, for example, in chemistry it is necessary to have reactors of different types [Vasilyev, Romanovsky, Yakhno, 1987], characterized by different concentrations of reactants, temperatures, geometry, etc. In contrast, an initially homogeneous metal sample, stretched at a constant speed, demonstrates the ability for spontaneous sequential generation of autowave modes of various types without any programmable rearrangement associated with external effects. This means that a deformable sample can be considered as a universal generator of autowave processes [Zuev, 2014].

3.4. The main characteristics of localized deformation autowaves

The phase autowaves of localized plastic flow, observed at the stages of easy slip and linear strain hardening, are the most convenient object for experimental research. A stable phase autowave pattern allows accurate measurement of its quantitative characteristics – the length λ and period T – using X–t diagrams.

The analysis and synthesis of numerous experimental data on the development of autowaves of localized plastic flow in various materials allowed us to establish the basic laws of their kinetics, which are important for understanding the nature of their occurrence. These include:

- dependence of the propagation speed of autowaves on the strain hardening coefficient;
- dependence of the autowave frequency on the wave number (dispersion relation;
- dependence of the autowave length on the sample length (scale effect);
- dependence of the autowave length on the parameters of the structure of the material, in particular, on the grain size (structural effect).

We will consider these patterns sequentially.

3.4.1. The speed of propagation of autowaves

As shown in Chapter 2, the experimentally determined propagation velocities of autowaves of localized plasticity are $10^{-5} \leq V_{aw} \leq 10^{-4}$ m/s. This value is related to the speed of movement of the moving grip of the testing machine by the ratio $V_{aw} \approx 10 V_{mach}$. A more in-depth analysis of the experimental data obtained under the condition of V_{mach} = const, showed that the propagation speed of localized plasticity autowaves depends on the strain hardening coefficient of the material

$$V_{aw}(\theta) = V_0^{(i)} + \frac{\Xi^{(i)}}{\theta}, \tag{3.18}$$

where the indices $i = 1$ and $i = 2$ refer to the stages of easyslip and linear strain hardening, respectively, and V_0 and Ξ are empirical

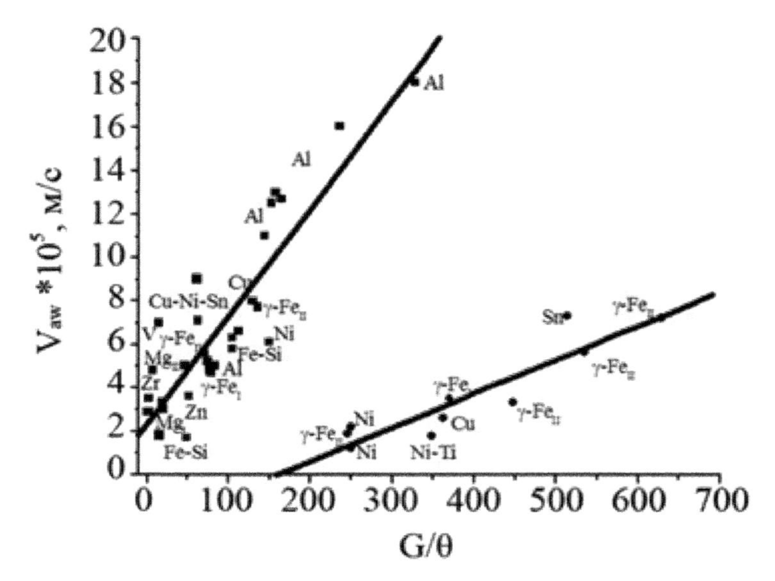

Fig. 3.8. The propagation velocity of phase autowaves as a function of strain hardening coefficient.

constants that differ for the mentioned stages of the plastic flow process. For the easy slip stage, $V_0^{(1)} = -2.5 \cdot$ 10–5 m/s, and $\Xi^{(1)} = 1.5\ 10^{-7}$ m/s, and for the stage of linear strain hardening, $V_0^{(2)} = 2.3\ 10^{-5}$ m/s, and $\Xi^{(2)} = 5.0\ 10^{-7}$ m/s. The dependence shown in Fig. 3.8, summarizes all the materials studied to date, the flow curve of which contains the indicated stages of strain hardening.

Let us consider some results on the form and nature of the dependence (3.18). In this relation, $V_0 \ll \Xi^{(1)}/\theta$ since $V_{aw} \ll \Xi^{(1)}/\theta$, that is, the order of magnitude of V_{aw} is determined by the coefficient Ξ, which has the dimension of velocity. We then discuss the meaning of the $V_t/\Xi \approx 10^{10}$ relation using the hypothesis of large Dirac numbers [1990], according to which large dimensionless ratios of physical quantities are not random, and to find out their physical meaning, it is useful to find the same order dimensionless ratio of other characteristics. In our case, they should be associated with deformation.

The values of the dynamic viscosity of the medium for the two deformation modes seem suitable for this purpose. The first of them corresponds to the quasi-viscous dislocation motion with a speed [Alshitz, Indenb, 1975a, b] $V_{disl} \approx (b/B) \times \sigma$, depending on the viscous drag coefficient $B \approx (1...3)\ 10^{-4}$ Pa·s. This value is determined by the interaction of moving dislocations with a phonon gas. The second, much larger viscosity of the medium is, according to ultrasonic measurements, $\eta \geq 3\times10^6$ Pa·s [Granato, Lyukke, 1969; Novik, Berry, 1975] and is due to the separation of dislocations from local barriers. Ostrovsky and Likhtman [1958] observed two levels of viscosity of a deformable medium, corresponding to low and high values of the flow stress, and an abrupt drop in this quantity with a change in the deformation mechanism. The ratio $\eta/B \approx 10^{10}$, therefore, based on the Dirac hypothesis, we can write down $V_t/\Xi \approx \eta/B$ or

$$V_{aw} \approx \frac{B}{\eta}\cdot\frac{1}{\theta}\cdot V_t. \tag{3.19}$$

The relation (3.19) is a combination of lattice (V_t) and deformation (B, η, θ) characteristics and seems promising for further analysis. Viscous stresses are defined as $\sigma_v = \mu \times dv/dy$, where μ is the dynamic viscosity and dv/dy is the velocity gradient v in the direction normal to the direction of wave x propagation. In this case, $\int_0^{l^*} \sigma_v dy = \int_0^{v} \mu dv$, where the integral on the left is the viscous resistance force acting on 1 border length of the moving acoustic or deformation (autowave) front. On the right, viscosity μ is consistently understood as η and B, and speed v, respectively, Ξ and V_t. Then $(\sigma_v l^*) = \eta\Xi$ and $(\sigma_v l^*) = BV_t$, where l^* is the thick"ness of the transition layer. The equality $\eta\Xi = BV_t$ in this case reflects the connection of processes in the elastic and plastic subsystems of a deformable medium; the possibility of recording it before the above analysis is not obvious, which confirms the heuristic value of using the hypothesis of large Dirac numbers.

The physical meaning of the dependence $V_{aw}(\theta)$ can also be analyzed within the framework of energy models of strain hardening [Nabarro, Bazinsky, Holt 1967; Mróz, Oliferuk, 2002; Oliferuk, Maj, Raniecti, 2004; Oliferuk, Maj, 2009]). In their opinion, the coefficient of strain hardening is the ratio of the energy of immobile dislocations accumulated during plastic deformation $W \approx Gb^2\rho_{stat}$ to the energy dissipated during the movement of dislocations $Q \approx \sigma b L_{disl} \approx \rho_m$. Here, W and Q are calculated for a unit volume; ρ_{stat} and ρ_m are the densities of accumulated and mobile dislocations, respectively, L_{disl} is their mean free path. Then

$$\theta \sim \frac{W}{Q} \sim \frac{Eb^2\rho_{stat}}{\sigma b L_{disl}\rho_m} \sim \frac{b}{\varepsilon_e L_{disl}}\cdot\frac{\rho_{stat}}{\rho_m}, \tag{3.20}$$

where $\varepsilon = \sigma/E$, and $L_{disl} \sim \sigma = \Lambda/\varepsilon_p$. In this case

$$V_{aw} \sim \Xi \cdot \frac{\Lambda}{b} \cdot \frac{\varepsilon_p}{\varepsilon_e} \cdot \frac{\rho_m}{\rho_{stat}} \tag{3.21}$$

In the case of a stationary plastic flow, an increase in the density of accumulated m/s efects causes an increase in W, which, according to equation (3.21), reduces V_{aw}. On the contrary, an increase in the energy dissipated into heat Q leads to the heating of the sample and increases the probability of thermally activated acts of plastic deformation, respectively, increasing V_{aw}.

The dependence $V_{aw} \sim \theta^{-1}$ can be explained under the assumption that $dV_{aw} \sim \mathbf{L}$, where $\mathbf{L} = \Lambda/\varepsilon - \varepsilon_2$ is the length of the slip trace at the stage of linear strain hardening. According to Seeger [1960], $\Lambda = \text{const}$, depending only on the type of n under investigation, and ε_2 is the deformation, corresponding to the beginning of the linear stage, where $\theta \approx \sqrt{nb/3\Lambda}$ and n is the number of dislocations in the cluster. Then $dV_{aw} \sim \mathbf{L} \sim \Lambda \sim 1/\theta^2$. For different materials $\varepsilon - \varepsilon_2 \approx \text{const}$ [Seeger, 1960; Berner, Kronmüller, 1969]. At the same time, the strain hardening coefficient θ is different for them, that is, $d\theta \neq 0$, and

$$dV_{aw} \sim \frac{nb}{(\varepsilon - \varepsilon_2)\theta^2} d\theta, \tag{3.22}$$

whence it follows that $V_{aw} \sim \theta^{-1}$.

To understand the meaning of formula (3.18) it is essential to realize that the dependence $V_{aw}(\theta^*)$, Fig. 3.9, where $\theta^* = d\sigma/d\varepsilon$ is the dimensional strain hardening coefficient at the stage of linear hardening of single crystals of the Fe_I alloy, has the form $V_{aw} = V_* + J/\theta^*$, where V_* 2.1 · 10^{-5} m/s and $J \approx 3.4 \cdot 10^4$ W/m^2. $J \approx \sigma V_{aw}$ has the meaning of the power of the flow of energy from the loading device, and $V_m = 10^{-6}$ m/s is the speed of movement of the moving jaw of the testing machine. If at the stage of linear strain hardening $\sigma \leq 10^3$ MPa, then for a given value of V_m, the value is $J \approx 10^4$ W/m^2, which is close to the above estimate. This means that the speed of autowaves is determined, among other things, by the characteristics of the loading device.

For this reason, the propaation speed of the autowave V_{aw} cannot be expressed through the material characteristics of the medium. This once again confirms that the discussed autowave processes of the localized plastic flow and the Kolsky plastic deformation waves [Kolsky, 1956] have a different nature and are not convertible to each other.

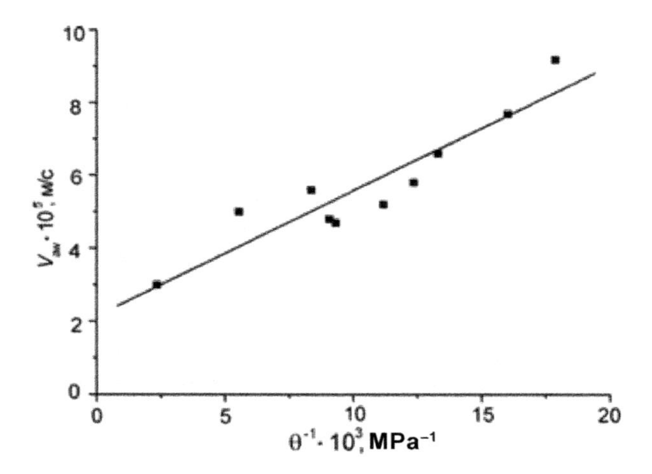

Fig. 3.9. Dependence of the velocity of plastic flow localization waves on the strain hardening coefficient at the stage of linear strain hardening of Fe_I single crystals

3.4.2. Dispersion of autowaves

An important characteristic for understanding the nature of autowave plastic flow localization processes is the form of the dispersion relation $\omega(k)$. The presence of dispersion indicates the existence of a set of spatial and/or time scales in the environment [Trubetskov, 2003]. At the stages of esy slip and linear strain hardening, when phase autowaves of localized plastic flow are formed, one can measure the length λ and period T of the autowave and obtain dispersion relations for Al polycrystals and single crystals of an Fe_I based alloy (linear strain hardening) and Cu, Sn, Fe_I single crystals (easy sliding), shown in Fig. 3.10 a. The dependences have the form

$$\omega(k)=\omega_0 \pm \alpha(k-k_0)^2, \tag{3.23}$$

where ω_0, k_0 and α are empirical constants depending on the type of material, the + sign refers to linear hardening, and the – sign refers to the stage of easy slip. Using the substitutions $\omega = \tilde{\omega}_0$ and $k = k_0 \pm \frac{\tilde{k}}{\sqrt{\text{sign}\alpha\cdot(\alpha/\omega_0)}}$, where $\tilde{\omega}$ and $\tilde{k}$ are dimensionless frequency and wave number, and sign $\alpha = \pm 1$ is a sign function, relation (3.24) is reduced to canonical dimensionless form

$$\tilde{\omega} = 1 + \tilde{k} \tag{3.24}$$

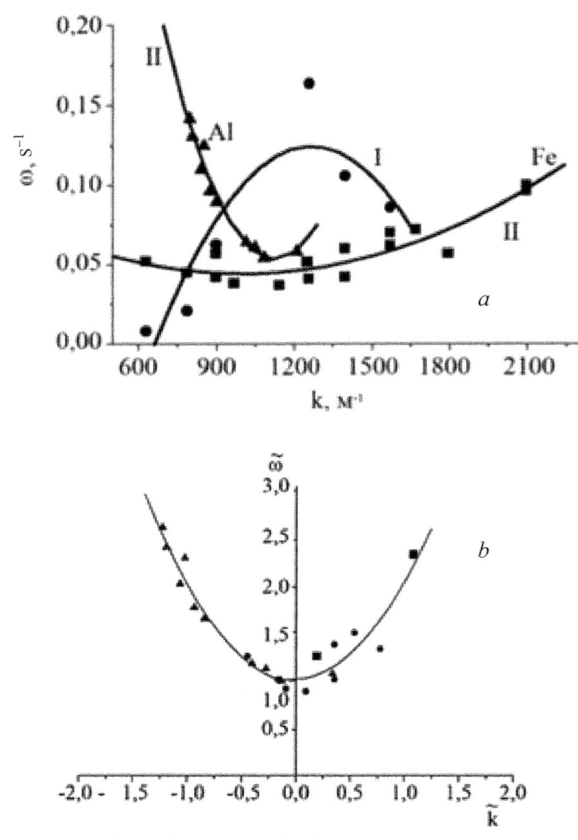

Fig. 3.10. Dispersion of localized plastic flow autowaves: initial data for Al and Fe (*a*); generalization of data in coordinates $\tilde{\omega}_0$ and $\tilde{k}_0$ (*b*).

and is presented in the form of a single parabola, as shown in Fig. 3.10, b.

The confirmation of the applicability of equation (3.24) for describing the dispersion of autowaves of a localized plastic flow is the linearization (Fig. 3.11) of the graph of Fig. 3.9, b in functional coordinates $\frac{\tilde{\omega}-\tilde{\omega}_0}{\tilde{k}-\tilde{k}_0}-\tilde{k}$. Here $\tilde{\omega}_0$ and $\tilde{k}_0$ are the coordinates of an arbitrary point of the dependence $\tilde{\omega}(\tilde{k})$.

The quadratic form of the dispersion relation is characteristic of the equations of nonlinear mechanics, describing self-organization processes in active nonlinear media, for example, for functions that satisfy the nonlinear Schrödinger equation [Dodd et al., 1988; Kosevich, Kovalev, 1989; Brown, Kivshar, 2008], often used for these purposes. The quadratic form of the dispersion relation is another proof that the localization of plastic flow is a process of self-organization of a deformable medium. Note that this form of the dispersion dependence for autowaves of the localized plastic flow (3.23) is similar to the dispersion law of de Broglie electron waves in the constant energy region, having the form $\omega = \frac{V}{\hbar} + \frac{\hbar}{2m}k^2 \sim k^2$ [Crawford, 1974].

In the work of Zuev, Khon and Barannikova [2010], an attempt was made to obtain the dispersion relation for autowaves of a localized plastic flow, based on the equation of the deformation of the medium, written in the form of the relation

$$\varepsilon_{ij}(r,t) = \varepsilon_{ij}^{e}(r,t) + \varepsilon_{ij}^{p}(r,t) + \varepsilon_{ij}^{ir}(r,t), \qquad (3.25)$$

where the first term on the right is elastic, the second is plastic, and the third is irreversible inelastic strain, r is the radius vector of the point in question.

Introducing the fractions of the volume occupied by the regions of reversible and irreversible strains and treating them as Landau order parameters [Landau, Lifshits, 2002] q and p, respectively, it was possible to get the quadratic dependence $\omega(k)$ in the form $\omega^2 = \omega_H^2 - ak^2 + bk^4$, where ω_H, a and b are constants. This corresponds to the experimental data shown in Fig. 3.12 in coordinates ω^2–k^2.

From the dispersion relation (3.24), the phase and group velocities of the process can be obtained [Crawford, 1974; Kosevich, Kovalev, 1989], and the phase of the autowave depends on k non-monotonously

$$\tilde{V}_{ph} = \frac{\tilde{\omega}}{\tilde{k}} \sim \frac{1}{\tilde{k}} + \tilde{k} \qquad (3.26)$$

and the group is linear in $\tilde{k}$.

$$\tilde{V}_{gr} = \frac{d\tilde{\omega}}{d\tilde{k}} \sim \tilde{k} \qquad (3.27)$$

The analysis of the dependences of $V_{ph}(k)$ and $Vgr(k)$ for localized deformation autowaves at the stages of easy slip and linear strain hardening provides important information about the nature of deformation processes. As can be seen from Fig. 3.13 a (stage of linear strain hardening), there is a value of the wave number at which the graphs $V_{ph}(k)$ and $V_{gr}(k)$ merge, that is, $V_{ph}=V_{gr}$. For this reason, the autowaves of the localized flow with a length of $\lambda < 5$ mm do not experience dispersion.

Table 3.2. Observed autowave modes

Stage (see Table 2.3)	Autowave mode	Materials
Yield plateau	Switching autowave	Fe–0,08%C, Fe–0,09%C–2%Mn–1%Si
Easy slip	Phase autowave	Cu, Ni, Zn, Fe–16%Cr–12% Ni–0,5%N, Cu–10%Ni–6%Sn; NiTi (equiatomic composition)
Linear strain hardening		Cu, Ni, Zn, Al, Ta, In, Co, Cd, Hf; Fe–13%Mn–1%C; Cu–10%Ni–6%Sn; Mg–2%Mn; Fe–3%Si
Paraboli strain hardening	Stationary dissipative structure	Cu, Ni, Co, Al, Ta, In, Cd, Hf; Fe–16%Cr–12% Ni–0,5%N; Cu–10%Ni–6%Sn; Ni3Mn (ordered); Mg–2%Mn; Fe–3%Si; NiTi (equiatomic composition)

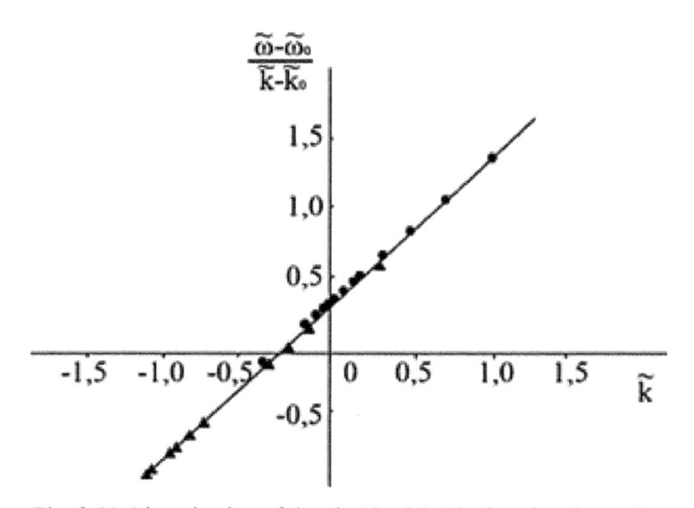

Fig. 3.11. Linearization of data in Fig. 3.9 *b* in functional coordinates.

Table 3.3. The rule of conformity of the hardening stages, patterns of localized plasticity and autowave modes

Plastic flow stages	Localized plasticity patterns	Autowave mode
Yield plateau	United moving front	Switching autowave
Easy slip, linear strain hardening	Set of moving fronts	Phase autowave
Parabolic strain hardening	Set of stationary fronts	Stationary dissipative structure
Prefracture	Merger of fronts with necking	Autowave collapse

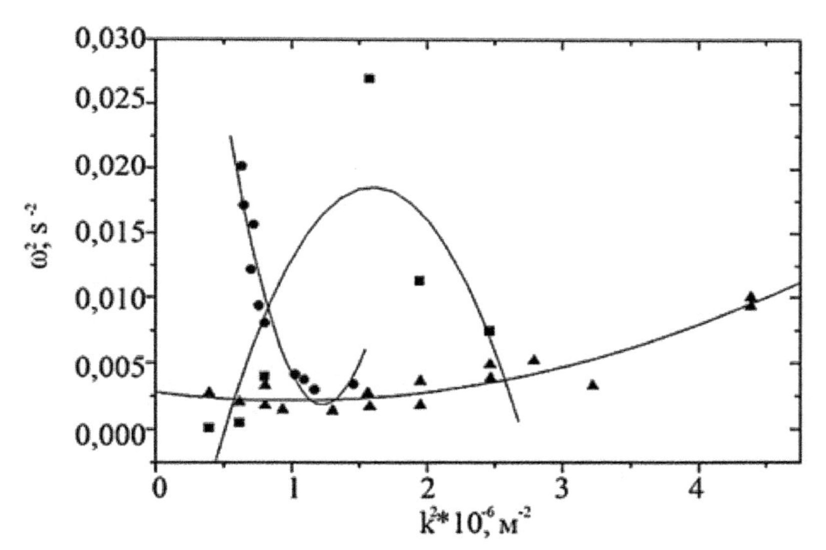

Fig. 3.12. Verification of the relation $\omega^2 = \omega_H^2 - ak^2 + bk^4$.

For the easy slip stage, as follows from Fig. 3.13, b, such a merger and, accordingly, the equality of the group and phase velocities of the autowave of the localized plastic flow is not achieved. This is the reason for the instability of the autowave process at this stage, which explains the well-known experimental difficulties associated with observing the stage of light slip [Berner, Kronmüller, 1969].

3.4.3. Scale effect with strain localization

The answer to the question about the influence of the size of

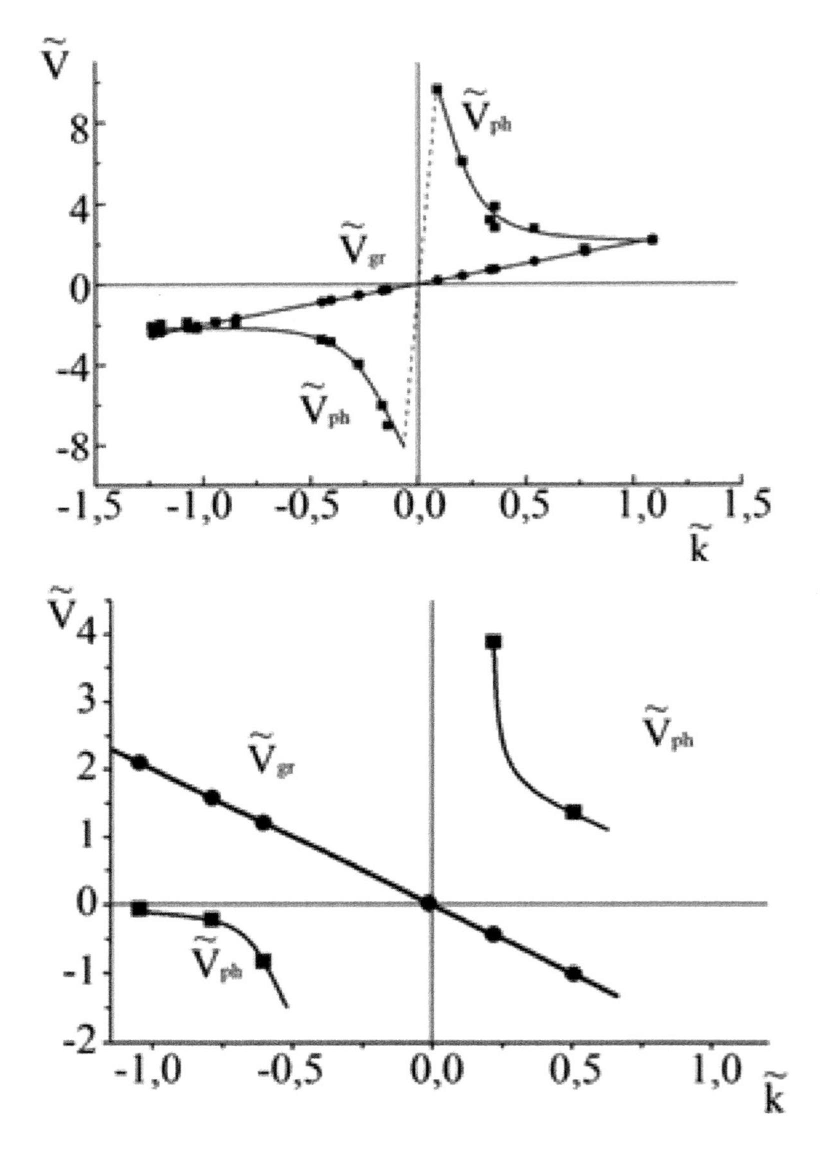

Fig. 3.13. Phase and group autowave speeds at the stages of linear strain hardening (*a*) and easy slip (*b*)

a deformed system on the parameters of localization of plastic deformation (scale effect) is important for understanding the nature of the phenomenon, since it provides additional information about the mechanisms of formation of autowaves of localized plasticity. The size effect was studied in the Zr–2.5 wt.% Nb alloy, whose

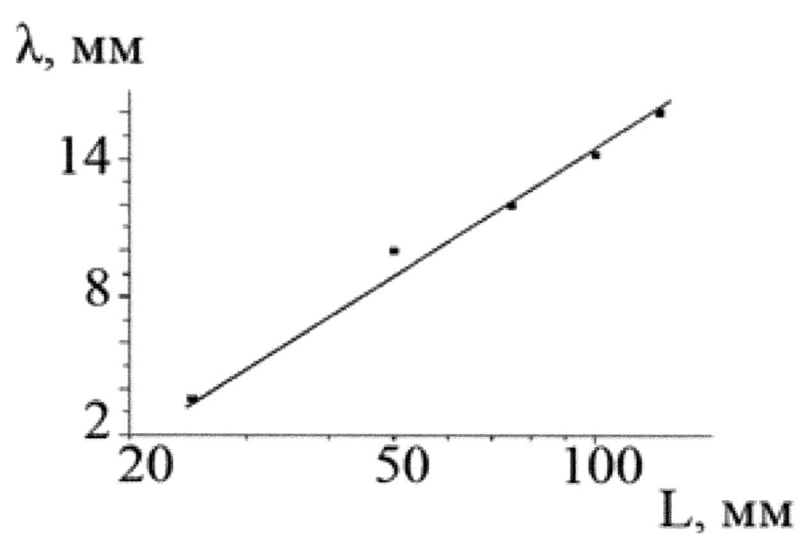

Fig. 3.14. The dependence of the autowave length on the length of the deformable sample

plastic flow diagram $\sigma(\varepsilon)$ has a stage of linear strain hardening [Zavodchikov, Zuev, Kotrekhov, 2014] on samples with lengths of 25, 50, 75, 100 and 125 mm, 5 mm wide and 1.6 mm thick.

It was established [Zuev et al. 2000] that the length of the autowave depends logarithmically on the sample size

$$\lambda(L) = \alpha \ln \frac{L}{L_0}, \tag{3.28}$$

as shown in Fig. 3.14. At the same time, $L_0 \approx 16$ mm, $\alpha \approx 7.8$ mm.

The obtained dependence can be interpreted assuming that the derivative $d\lambda/dL > 0$ is inversely proportional to the probability of nucleation of the localization centre p. If the latter, in turn, is proportional to the sample length, then $d\lambda/dL \sim 1/p \sim 1/L$. Thus, $d\lambda = \alpha\, dL/L$, whence equation (3.28) follows, the coefficient α in which has the meaning of the scale of spatial non-uniformity of plastic deformation in the sample.

From equation (3.28) it follows that $\lambda = 0$ at $L_0 \approx 2\alpha \approx 16$ mm, so that the value of L_0 should be considered as the minimum sample size in which the occurrence of autowaves of plastic flow localization is possible. In samples of length $L \leq L_0$ under tension one should

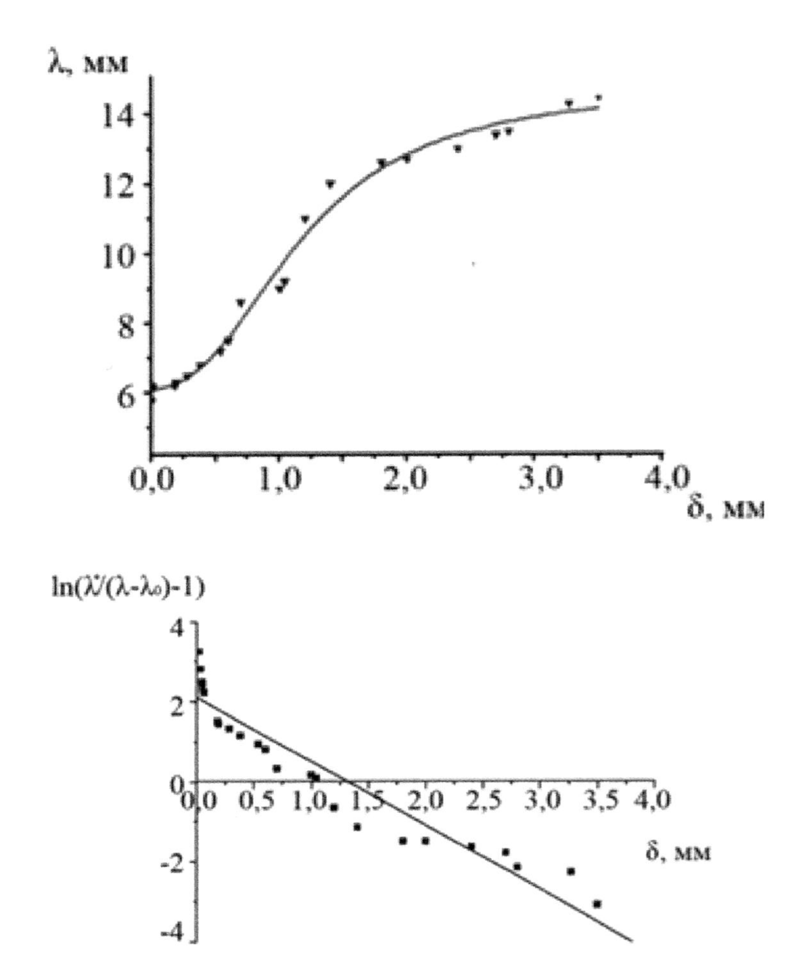

Fig. 3.15. The dependence of the length of the autowave of localization of plastic deformation lfrom the grain size dfor Al: initial data (a); same data in functional coordinates (b)

expect a uniform distribution of strain. Indeed, when testing a sample of the same alloy 18 mm long, no macroscopic localization of plastic flow was detected under the same deformation conditions.

The existence of such a threshold explains why in the literature devoted to the problems of strain localization, is rarely mentioned in the observation of periodic distributions of plastic strain localization. The reason is that it is often used for testing samples of small length, in which this effect is impossible. The localization of deformation in large specimens seems to be the reason for the existence of a scale effect (dependence of strength on size [Bogachev, Weinstein, Volkov, 1972]), which plays an important role in engineering.

3.4.4. Autowave parameters and material structure

The question of the dependence of the autowave length of localization on the grain size was investigated on polycrystalline aluminium with a purity of 99.85 wt.% [Zuev, Semukhin, Zarikovskaya, 2002; Zuev, Zarikovskaya, Fedosova, 2010; Zuev, et al., 2013]. Using the method of post-deformation recrystallization [Schmid, Boas, 1938] at a temperature of 853 K grains of size $5 \cdot 10^{-3} \le \delta \le 10$ mm were grown in samples. The dependence of the strain localization wavelength on the grain size $\lambda(\delta)$ established in these experiments is shown in Fig. 3.15 a.

The analytical form of the function shown in Fig. 3.15 a can be obtained from the following considerations. Suppose that with increasing grain size, the wavelength of localized deformation also increases due to the elongation of the shear band during the deformation event. We take into account that when the grain size approaches the sample diameter, the growth rate should slow down, since in this case the length of the slip band remains less than the grain diameter. Taking this into account, we write the differential equation relating the quantities λ and δ in the form

$$\frac{d\lambda}{d\delta} = a\lambda - a^*\lambda^2. \tag{3.29}$$

On the right side of equation (3.29), $a > 0$ and $a^* > 0$ are dimensional constants; the quadratic term $a^*\lambda^2$ takes into account the slowing down of the increment λ_B in the region of large δ. Integrating equation (3.29) leads to the well-known logistic curve equation [Volterra, 1976]

$$\lambda = \lambda_0 + \frac{\lambda^*}{1 + C \cdot \exp(-a\delta)}, \tag{3.30}$$

where $\lambda_0 \approx$ 4 mm, a = 1.4 mm^{-1}, a^* = 8.8×10^{-2} mm^{-2}, λ^* = $a/a^* \approx 16$ mm, and $C \approx$ 2.25 - a dimensionless integration constant . By representing the dependence $\lambda(\delta)$ in functional coordinates $\ln\left(\frac{\lambda^* - \lambda_0}{\lambda} - 1\right) - \delta$ we can verify the applicability of equation (3.30) in the studied range of values (Fig. 3.15, b).

Consider the limiting cases of dependence (3.30). In the region of small grains with $\delta \leq 5\times10^{-2}$ m, obviously, $a^*\lambda^2 \ll a\lambda$ and $d\lambda/d\delta \approx a\lambda$, which, when substituted in (3.29), results in $\lambda \sim \exp\delta$. In the range of grain sizes $5\times10^{-2} < \delta < 2.5$ mm, where the wavelength growth slows down, it can be assumed that the relative gain of λ is proportional to the number of grains on the working length of the sample L, that is, $d\lambda/d\delta \sim L/\delta$ or $d\lambda \sim L(d\delta/\delta)$. Hence, obviously, the logarithmic relation characteristic of coarse-grained Al is $\lambda \sim \ln\delta$.

The change in the dependence of $\lambda(\delta)$ from exponential to logarithmic occurs under the condition $\lambda = \lambda^*/2$, which corresponds to 0.25 mm. Thus, the length of the autowave of a localized plastic flow is a structurally sensitive value, and depends in a complex way on the grain size of the deformable material.

It is known that the characteristics of strength and plasticity of materials are structurally sensitive properties and depend on the grain size. Therefore, it seems possible to have a deeper connection between the autowave characteristics of the localized plastic flow and the grain size. It is convenient to analyze this problem using the Hall–Petch relation [Friedel, 1967; Honeycomb, 1972; Kovács, Ching, Kovács Csetènyi. 2002; Nabarro, 2004; Counts et al., 2008; Pelleg, 2013; Lim et al, 2014; Movchan, Firstov, Lugovskoy, 2015], recorded for the flow stress σ_f, determined under the condition of permanent deformation

$$\sigma_f = \sigma_0 + k_f\delta^{-1/2}. \tag{3.31}$$

The parameters σ_0 and k_f are related to the metal structure. Therefore, we will further present the properties of the metal under investigation as a function of $\delta^{-1/2}$.

The dependence of the autowave length on the grain size, presented in the coordinates λ– $\delta^{-1/2}$, as shown in Fig. 3.16, has a kink at $\delta \approx 0.25$ mm (at the inflection point of the function $\lambda(\delta)$, that is, under the condition $d^2\lambda/d\delta^2 = 0$).

In addition, in the investigated grain size range of $5\cdot10^{-3} \leq \delta \leq 10$ mm, an abrupt change in the dependence of the flow stress on the grain size $\sigma_f(\delta)$ was found at $\delta \approx 0.10...0.12$ mm, as shown in Fig. 3.17. A similar jump in the dependence of the elastic limit of pure Al on the grain size was observed earlier by Jaoul [Jaoul, 1957] at $\delta \approx 0.2$ mm.

In this case, the principal difference in the behaviour of the parameters σ_0 and k_f of equation (3.30) at $\delta > 0.1$ mm and $\delta < 0.1$

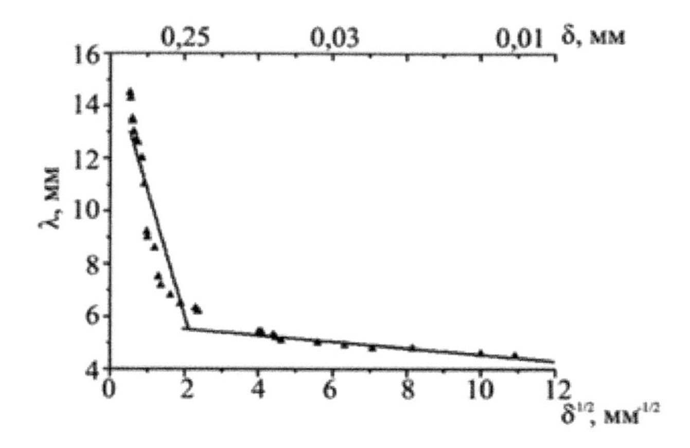

Fig. 3.16. The dependence of the autowave length of a localized plastic flow on the grain size in the coordinates $\lambda - \delta^{-1/2}$.

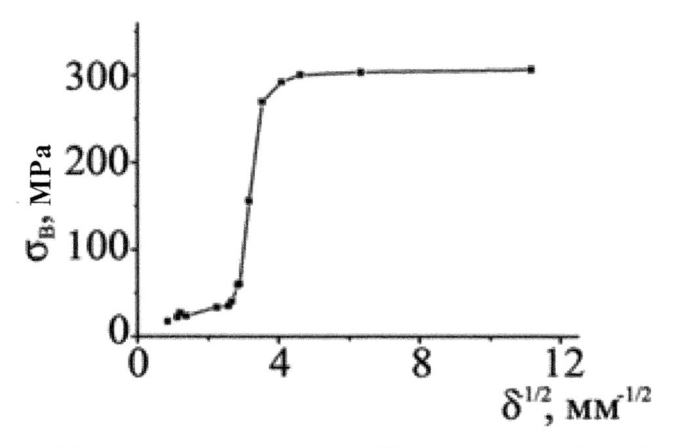

Fig. 3.17. Ultimate strength of polycrystalline Al as a function of grain size.

mm [Zuev, Zarikovskaya, Fedosova, 2010; Zuev, et al., 2013] is of special interest. As follows from Fig. 3.17, the proportionality of $\delta_f \sim \delta^{-1/2}$ is fulfilled in all cases, but at the boundary value of the grain size 0.1 mm, the character of the dependences changes.

From Fig. 3.18 it is clear that:

- in the range of $\delta > 0.1$ mm, the dependence graphs (3.31) $\sigma_f = \sigma_0 + k_f \delta^{-1/2}$ form a set of straight lines, which corresponds to $\sigma_0 = \text{const}$ and $k_f \sim \exp(\varepsilon/\varepsilon_c)$;
- in the range of $\delta < 0.1$ mm, the dependency plots (3.31) $\sigma_f = \sigma_0 + k_f \delta^{-1/2}$ are a family of parallel straight lines for which $k_f = \text{const}$ and $\sigma_0 \sim \exp(\varepsilon/\varepsilon_c)$.

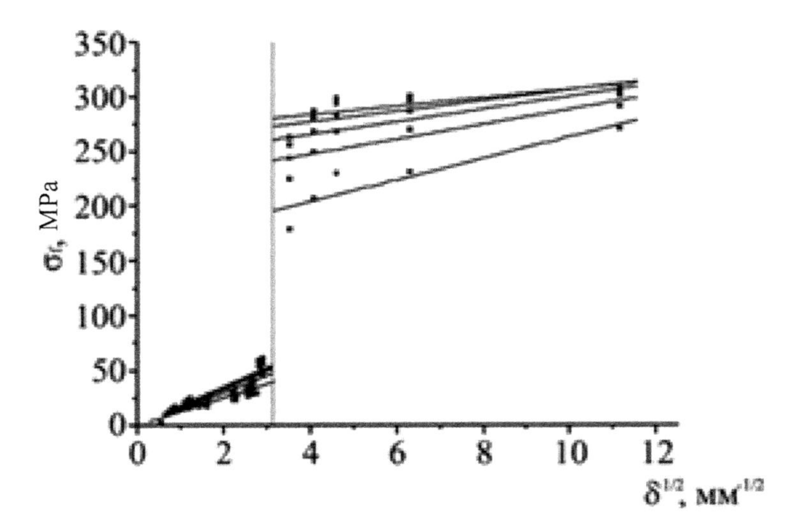

Fig. 3.18. Hall–Petch relationship for polycrystalline Al.

Within the accuracy of the experiment in both cases, $\varepsilon \approx 4.5 \cdot 10^{-2}$. We can assume that the different behavioir of the parameters σ_0 and k_f in the intervals $\delta > \delta_b$ and $\delta < \sigma_b$ is responsible for the difference in the character of the Hall–Petch proportionality $\sigma_f \sim \delta^{-1/2}$ in different parts of the investigated grain size range $8 \cdot 10\text{–}3 \leq \delta \leq 5$ mm of polycrystalline aluminium.

Thus, the parameters σ_0 and k_f of the grains depend on the grain size δ and strain ε. In addition, Fig. 3.18 shows that there is a boundary grain size of $\sigma \approx 0.1$ mm, at the transition of which the flow stress σ_f increases abruptly. The change in the Hall–Petch dependence corresponds to that shown in Fig. 3.17 more than threefold increase in ultimate strength σ_B when moving from a range of large to a range of small grains.

We now turn to an analysis of the behaviour of the autowave mode of a localized plastic flow in tests of coarse-grained ($\delta > 0.4$ mm) Al. The previously established pattern, according to which at the stage of linear strain hardening the centres of localized plasticity move with a constant velocity $V_{aw} \approx (10^{-5}\text{–}10^{-4})$ m/s, is also satisfied in the case studied. The localization pattern is a phase autowave of plastic flow with a length λ and a propagation velocity $V_{aw} = V_0 + \Xi/\theta$. The dependence $V_{aw}(q)$ is divided, as shown in Fig. 3.19 a, into two parts.

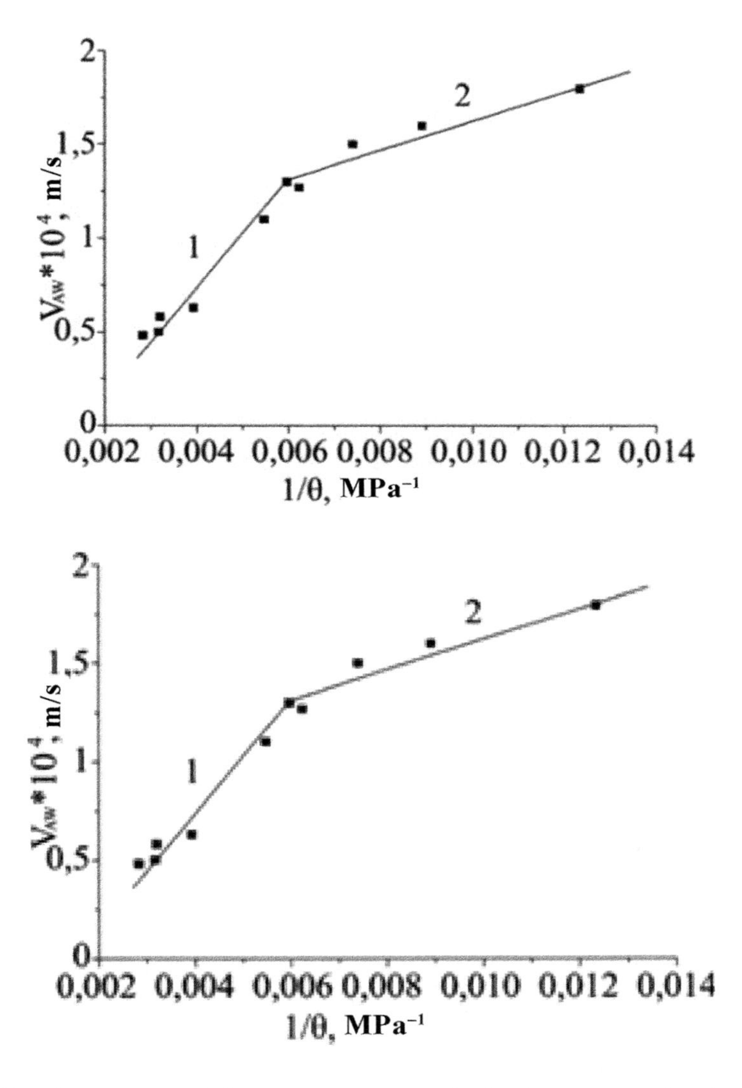

Fig. 3.19. The dependence of the propagation speed of autowaves on the strain hardening coefficient (*a*) and on the grain size (*b*). Grain size ranges 1 – $0.005 \leq \delta \leq 0.15$ mm; 2 – $0.15 \leq \delta \leq 5$ mm.

Since the strain hardening coefficient of Al polycrystals changes with varying grain size and can be obtained from the initial flow curves $\sigma(\varepsilon)$, it is possible to transform the graph of Fig. 3.17 to the the graph in coordinates V_{aw}–$\delta^{-1/2}$, as it is done in Fig. 3.19 b. It is clear from the graph that

$$V_{aw} = V_f - \vartheta\delta^{-1/2} \tag{3.32}$$

moreover, the slope of these dependences changes at $\delta \approx 0.1...0.12$ mm, but their linearity in size $\delta^{-1/2}$, characteristic of the Hall–Petch relation [Pelleg, 2013], is satisfied.

The dispersion ratio for polycrystalline aluminum is also associated with the grain size. As follows from Fig. 3.20 a, the points on the graph of the dependence $\omega(k)$ are located in accordance with the grain size. In this case, the minimum of the dispersion curve $\omega(k)$ corresponds to a size of $\delta \approx 0.1$ mm.

The phase $V_{ph} = \omega/k$ and group $V_{gr} = d\omega/dk$ autowave speeds of localized plastic deformation are functions of the grain size. The dependences $V_{ph}(\delta)$ and $V_{gr}(\delta)$, shown in Fig. 3.20 b, intersect at the point corresponding to the minimum of the function $\omega(k)$, that is, at $\delta \approx 0.1$–0.12 mm.

Finally, a change in the properties of polycrystals upon reaching the boundary value of $\delta \sim 0.1$ mm was confirmed by measuring the ultrasonic propagation velocity by the method of autocirculation of ultrasonic pulses [Muravyov, Zuev, Komarov, 1996] at a frequency of 2 MHz in Al polycrystals over the entire grain size range under study. As shown in the picture. 3.20, at $\delta \approx 0.1$ mm, this value drops abruptly. Earlier, this dependence was analyzed for a grain size interval of $\delta > 0.1$ mm in [Zuev et al., 2000], where it was shown that it looks like $V_t = V_{in} - \beta\delta^{-1/2}$.

The decrease in the propagation velocity of ultrasound during grain refinement is associated with the scattering of ultrasonic waves on structural inhomogeneities of the medium, whose role is played by grain boundaries. In the present work, having expanded the range of grain sizes to the side of fine-grained structures, it was possible to establish that when going from $\delta > 0.1$ mm to $\delta < 0.1$ mm, V_t drops sharply, as shown in Fig. 3.21.

Summarizing the results obtained, it should be noted that all of them definitely indicate the existence of a boundary grain size $\delta = \delta_b \approx 0.1$ mm in aluminum polycrystals. This boundary value of the grain size corresponds to a stepwise change in both the autowave characteristics and the mechanical properties of the metal, as shown in Table 3.4. It can be assumed that the existence of the threshold value of the grain size is associated with a change in the state of the grain boundaries in the metal [Zuev et al., 2013].Patterns of localized plastic flow are autowave modes arising during plastic deformation

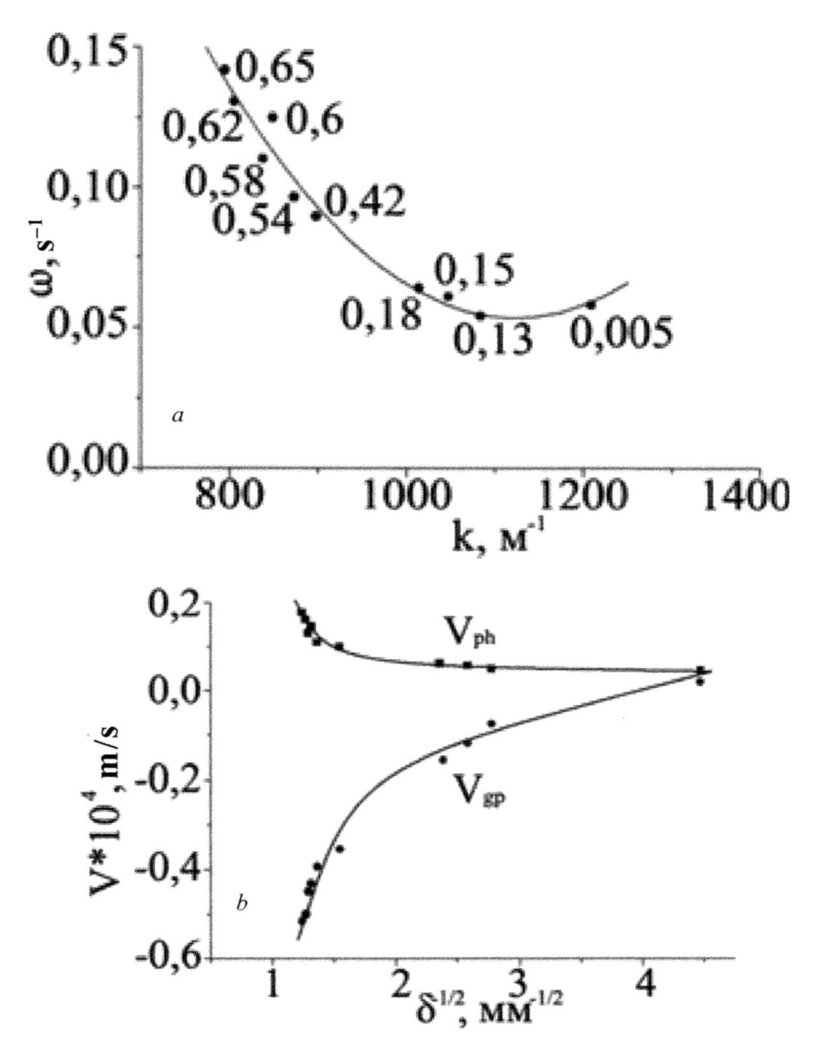

Fig. 3.20. Autowave dispersion in polycrystalline Al. Figures at the points – grain size in millimeters. Minimum dependence corresponds to 0.1 mm (a); phase and group velocity as a function of grain size (b).

of the medium. In this case, each pattern of localized deformation is assigned a certain autowave mode.

The equations proposed for describing the autowave processes of a localized plastic flow and their analysis made it possible to establish a connection between the developed concepts and the theory of dislocations and to understand the laws of the plasticity

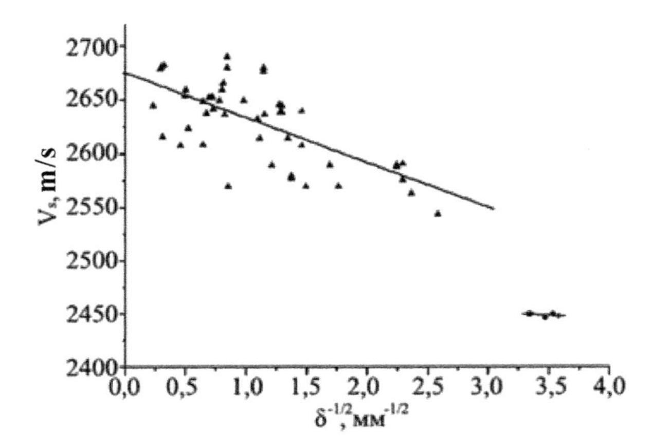

Fig. 3.21. Experimental data on the propagation speed of ultrasound in polycrystalline Al, as a function of grain size.

phenomenon. In particular, it was possible to explain the propagation speed of autowaves, the shape of their dispersion ratio, and the influence of the material structure on the autowave parameters of localized plasticity.

The autowave model for the development of localized plasticity made it possible to accurately and consistently describe plastic deformation as a collective effect. This idea, expressed by Seeger and Frank [1987] and in a more general form, by Prigogine [1985] and Manevich [2007], was embodied in the observation and explanation of autowave deformation modes corresponding to certain stages of strain hardening.

Due to the introduction of localized plasticity autowaves, the theory of plasticity acquires a structure similar to the structure of the theory of elasticity. In the latter, the redistribution of elastic stresses occurs due to the propagation of elastic waves. In the course of plastic flow, the redistribution of deformations is also determined by the wave process – localization autowaves. This allows one to naturally introduce the time factor in the description of the processes of plastic deformation.

And, finally, the use of autowave representations made it possible to present the plastic flow as the result of a regular evolution of the autowave structure of a deformable medium. It turned out that a deformable sample with an initially homogeneous structure can work as a universal generator of autowave modes of various types.

Table 3.4. Critical grain size for autowave parameters and mechanical properties of polycrystalline aluminium

	Characteristic	Critical grain size δ_b, mm
1	Ultimate strength, σ_B	0.15
2	Discontinuty of the Hall–Petch relation, $\sigma_f = \sigma_0 + k_f\delta^{-1/2}$	0.1
3	Change of the coefficient of the dependence of the velocity of the autowaves on the strain hardening coefficient in the relation $V_{aw} = V_0 + \Xi/\theta$	0.1...0.15
4	Intersection of the dependences $V_{ph}(\delta)$ and $V_{gr}(\delta)$	0.1...0.15
5	Speed of ultrasound V_S	0.1
6	Minimum on the dispersion curve $\omega(k)$	0.1...0.15
7	Inflection point on the dependence $\lambda(\delta)$	0.25

4

Two-component plastic flow model

4.1. On the principles of plastic flow model construction

To create a model of development of a localized plastic flow, first of all, it is necessary to clearly present its structure and principles of operation. It is quite obvious that the model should be based on the fundamental ideas about the self-organization of the defect structure of a deformable medium, originally formulated by Seeger and Frank [1987]. This statement actually determines that theoretical principles for the development of a model should be borrowed from synergetics, which, in fact, is the theory of structure formation [De Groot, Mazur, 1964; Haken, 1985, 2014; Nicolis, Prigogine, 1979; Cross, Hohenberg, 1993; Prigogine, 2005].

A synergistic approach to the construction of a plasticity model implies a conscious rejection of attempts to take into account the contribution of all microscopic plasticity acts and the transition to the description of specific collective movements in a deformable medium [Prigogine, 1985; Manevich, 2007], for which it is convenient to consider the autowave modes of localized plasticity, introduced in Chapter 3. At the same time, the synergetic approach requires

- *approximate nature* of the model, that is, its insensitivity to small changes in factors that can influence the process being mode;led [Chernavsky, 2004],
- taking into account the *nonlinear* properties of the environment, making it suitable for structure formation [Scott, 2007; Mishchenko et al., 2010].
- the model should describe the interaction of at least two components

(factors) of the deformation process, which can be considered as *plastic deformation and elastic stresses* (elastic strains). The basis for the use of two interacting factors are the following considerations:

- the development of autowave processes in deformable media is controlled by the competition of two factors and described by the system of two equations (3.3) for deformations and (3.4) for stresses,
- deformation in the general case is represented as the sum of two components – elastic and plastic [Hill, 1956],
- the coefficient of strain hardening in the theory of Nabarro, Bazinsky, Holt [1967] is determined by the ratio of two energies – the energy of defects stored in the deformed medium and the energy scattered into heat,
- The developed model should provide:
- the use of precisely defined concepts that are used in the physics of plasticity,
- the adequacy of the mathematical formulation of the model to the experimental results obtained in our research,
- the possibility of reducing to it the dislocation model of plasticity, which is used to describe the hardening of crystals.

4.2. Construction of a two-component plasticity model

Formation of autowave modes in open physical [Sobolev, 1991; Olemskoi, Sklyar, 1992; Olemskoi, 2009], chemical [Nicolis, Prigogine, 1979; Polak, Mikhailov, 1983; Barelko et al., 1993; Lavrova, Postnikov, Romanovsky, 2009] or biological [Romanovsky, Stepanova, Chernavsky, 1975, 1984; Schierwagen, 1991; Lobanov, Kurylenko, Ukrainets, 2009; Ivanitsky, 2017] systems are currently considered as the result of self-organization of active nonequilibrium media [Kerner, Osipov, 1990]. When discussing this problem, Kadomtsev [1997] proposed a fruitful concept, according to which: "*In complex systems with a complexly organized structure, it is possible to split a single system into two closely related subsystems. We can call one of them dynamic or power, and the second can be called the information or control subsystem. Those structural elements that can strongly influence the dynamics of the system by relatively small perturbations (signals) are naturally placed into the control structure. Thus, complex systems themselves can stratify into two levels of hierarchy.''* This concept, which is in line with modern ideas about the self-organization of active media [Ebeling, 1979; Cross, Hohenberg, 1993] seems promising to explain the nature of plastic flow autowaves.

4.2.1. Two-component model: structure and operation

The structure of the model being developed is based on the idea that the theoretical description of autowave processes, regardless of their nature, as shown by Ebeling, [1979], Hakken [1985], Vasilyev, Romanovsky, Yakhno [1987], is based on the consideration of autocatalytic competition and damping factors. This situation has already been discussed above in the analysis of the autowave mechanisms of localized plastic flow and the factors controlling the process of plastic flow. It can be assumed that the dynamic and signal subsystems, whose existence is postulated by Kadomtsev [1997], are closely related to these factors.

Given the specificity of a plastically deformable medium, first of all, it is necessary to determine the nature of the signal and dynamic subsystems in it. For this purpose, we formulate almost obvious mandatory requirements for these subsystems:

- subsystems should be determined by phenomena controlling the plastic process;
- between subsystems there should be an interaction of such power that the acts in one subsystem caused a response in another;
- the nature of the causal relationship between the phenomena characteristic of each subsystem should be strictly defined, and the mechanism of interaction between the subsystems should allow the possibility of checking and quantifying the characteristic parameters of plastic flow.

Formally, in the case of plastic deformation, the stratification of a deformable system can be represented by the method shown in Fig. 4.1.

The requirements listed above are fulfilled within the framework proposed in our works [Zuev, 1994, 1996; Zuev Danilov, Gorbatenko, 1995; Zuev, Danilov, 1998a, b; 1999] a two-component model that uses ideas about the relationship and interaction of the elastic and plastic components of deformation. The development of deformation acts in this model is determined by the mosaic shown in Fig. 4.2

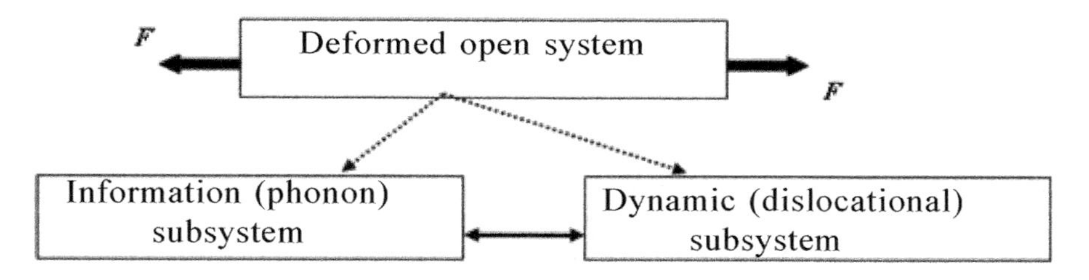

Fig. 4.1. Block diagram of a two-component plastic flow model.

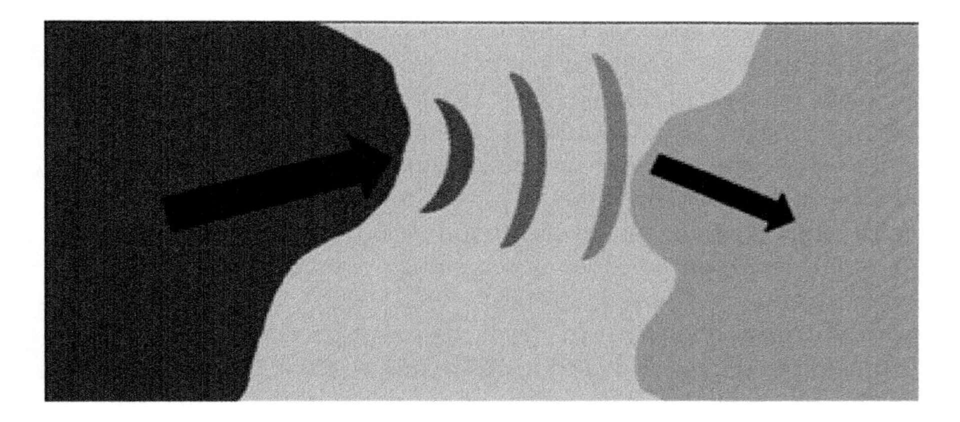

Fig. 4.2. Diagram of the mechanism of sequential activation of stress concentrators during deformation.

of deformed and strained volumes of a medium to varying degrees [Roitburd, 1974]. The mosaic of the deformation field is associated with a system of stress concentrators of various scales, which can be in a waiting or active (relaxing) state. Relaxation (decay) of a concentrator in such a model is considered as an act of plastic flow [Dotsenko, Landau, Pustovalov, 1987; Suzuki, Yoshinaga, Takeuchi, 1989]

Within this model, subsystems are defined as follows:

- the dynamic subsystem is a combination of dislocation shifts, twins, and other acts of plastic deformation, directly leading to plastic forming [Suzuki, Yoshinaga, Takeuchi, 1989],

- the signal subsystem is a set of elastic pulses of acoustic emission generated by plasticity acts and initiating such acts [Boyko, Natsik, 1978; Williams, 1980; Boyko, Garber, Kosevich, 1991, Skvortsov, Litvinenko, 2002; Blagoveshchenskii, Panin, 2017].

The specific scenario of the functioning of the two-component model is as follows. Each relaxation act in the dynamic subsystem is accompanied by the emission of an acoustic pulse. Such a pulse is absorbed by another hub and induces its relaxation, which is equivalent to an event in the dynamic subsystem. The behaviour of a deformable medium is characterized by the fact that random elastic pulses roam in the system of elastic stress concentrators the imposition of which on the static fields of the concentrators increases the probability of the implementation of relaxation acts of plastic deformation. It is easy to see that the two-component model has signs of Einstein's theory of thermal radiation (Rumer, Ryvkin, 2000) and combines the effects of acoustic emission [Williams, 1980] and acoustic induced plasticity effect [Kozlov, Mordyuk, Selitser, 1986; Malygin, 2000], usually studied separately.

In the general case, the implementation of the described interaction mechanism becomes possible due to the nonlinearity of the deformable medium. In general, nonlinearity suggests the possibility of generating harmonics when a complex signal propagates in such an environment [Pantel, Putthof, 1972]. This suggests a special type of connection between the factors that determine the behavior of the environment [Knyazeva, Kurdyumov, 1994; Scott, 2007] due to the corresponding harmonics. In our case, if during the decay of the stress concentrator A (Fig. 4.3) an acoustic signal is generated that contains a harmonic of a certain frequency, then the spectrum of the receiving concentrator B (Fig. 4.3) of a similar structure has the ability to absorb such a harmonic with a corresponding increase in the effect. In fact, this possibility was considered earlier in the work of Khon et al. [2008] (see section 3.2.3).

4.2.2. Numerical estimates of the capabilities of the model

In accordance with the proposed model, plastic deformation in the form of generating various autowave modes is realized as a result of the interaction of the dynamic and signal subsystems. The mechanism of the two-component model, shown schematically in Fig. 4.3, includes three successive stages:

- plastic shear (relaxation decay of the stress concentrator in the dynamic subsystem), which generates an acoustic emission pulse (point A in Fig. 4.3);
- the absorption of the energy of an acoustic pulse by another stress concentrator with the initiation of its thermally activated relaxation, that

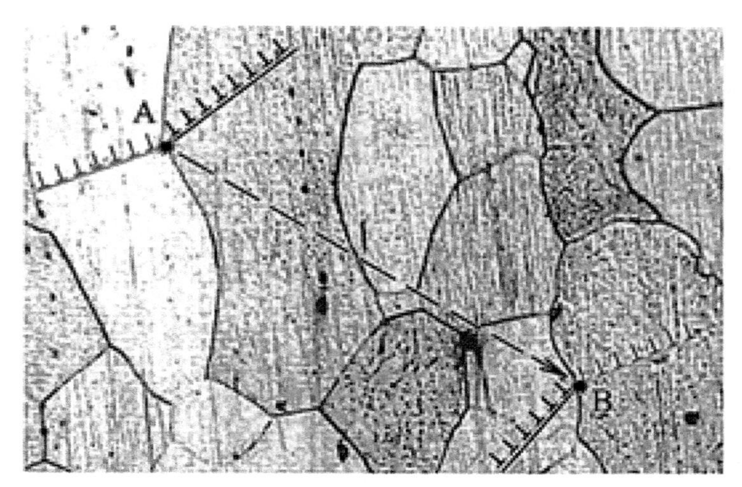

Fig. 4.3. The mechanism of functioning of the two-component model. The initial stress concentrators are indicated by bold ⊥. The resulting shifts are indicated by thin ⊥. Dashed arrow AB – acoustic pulse propagation path.

is, a new shear process (point B in Fig. 4.3) [Friedländer, 1962; Regel, Slutsker, Tomashevsky, 1974; Malygin, 2000];

- acoustic emission when implementing a new shift with repeating the above steps.

The credibility of the proposed acoustic mechanism, in principle, is confirmed by the well-known Wallner effect – the appearance of relief grooves on the cleaved surface of brittle bodies (Fig. 4.4, a) [Kerkhof, 1971]. They occur in accordance with the scheme shown in Fig. 4.4 b, due to the curvature of the crack front under the action of ultrasound pulses with a frequency of ~10 MHz emitted by a crack during its growth [Drozdovsky, Fridman, 1960; Kerkhof, 1971].

According to the scheme of geometric analysis of the effect shown in Fig. 4.4 b the crack originates at point *B*. The successive positions of its front are shown by solid lines 1...8. When the destruction front F_0 reaches any defect *S* on the surface, the latter begins to emit elastic waves, shown in the diagram by dashed lines 1...3. The encounter of the fracture front F_A with the elastic wave occurs at point *A* on line *L*, along which the front deviates from the original plane of destruction.

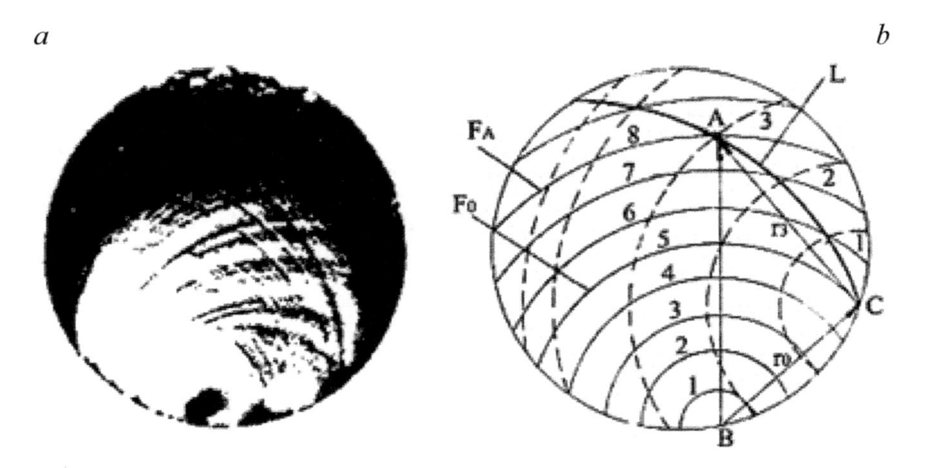

Fig. 4.4. The type and origin of the Wallner lines: fractogram [Kerkhof, 1971] (*a*); scheme of appearance [Drozdovsky, Friedman, 1960] (*b*).

For energy analysis of the problem of crack deviation, it is possible to estimate from below the order of magnitude of the energy of such pulses, assuming that it is spent only on increasing the surface area of fracture ΔA, and neglecting other dissipation channels. In accordance with the data of Fig. 4.4, $\Delta A \approx 10^{-8}$ m^2 with a depth of striations on the fracture surface of ~1 μm and a sample diameter of ~10^{-2} m. With the density of the surface energy characteristic for metals $\gamma \approx 1$ J/m^2 [Adamson, 1979] pulse energy, which bends the crack front, $\Delta W \geq \gamma \Delta A \approx 10^{-8}$ $J = 6.25 \ \ 10^{11}$ eV, that is, such pulses in solids can have sufficiently high energy.

This energy is spent on the formation of new dislocation lines with a total length of $\gamma \Delta S/Gb^2 \approx 10^3$ m. With the characteristic length of the dislocation loop emitted by the Frank-Reed source of ~50 μm [Van Buren, 1962; Friedel, 1967], this corresponds to ~$5 \cdot 10^6$ loops. The source is locked with a reverse stress, emitting ~50 loops, so that the energy of the acoustic pulse is enough to activate ~10^5 sources and generate new shifts. If the characteristic distance between sources at the stage of linear strain hardening is ~$4 \cdot 10^{-6}$ m [Berner, Kronmüller, 1969; Caillard, 2010], almost all sources in the sample volume can be activated by acoustic pulses.

Now we will conduct a quantitative assessment of the applicability of the proposed model for the development of a localized flow, for which purpose we will quantify the validity of the proposed model by comparing the waiting times for thermally activated acts of shear relaxation [Osipov, 1962, 1986; Engelke, 1973] in the absence of an acoustic impulse

$$\vartheta_{ab} \approx \omega_D^{-1} \exp\left(\frac{U_0 - \gamma\sigma}{k_B T}\right), \tag{4.1}$$

and in its presence

$$\vartheta_{ap} \approx \omega_D^{-1} \exp\left[\frac{U_0 - \gamma\sigma - \delta U_{ac}}{k_B T}\right] \approx \omega_D^{-1} \exp\left[\frac{U_0 - \gamma(\sigma + \varepsilon_{ac} E)}{k_B T}\right]. \tag{4.2}$$

In these ratios, U_0 is the height of the potential barrier overcome when a relaxation act develops, $\gamma \approx b^2 l \approx 10^4 b^3$ the activation volume of such an act. For calculations using the formulas (4.1) and (4.2), we assume that the activation enthalpy of the process is $U_0 - \gamma\sigma \approx 0.5$ eV [Regel', Slutsker, Tomashevsky, 1974]. An acoustic pulse with an amplitude of elastic strain ε_{ec} reduces this value by $\delta U \gamma \varepsilon_{ec} E \approx 0.1$ eV. The calculation for $k_B T \approx 1/40$ eV shows that $\vartheta_{ab} \approx 5 \cdot 10^{-5}$ s and $\vartheta_{ab} \approx 9 \cdot 10^{-7}$ s $<< \vartheta_{ab}$. Even with obvious approximation, this assessment confirms the validity of the model, explaining the acceleration of plastic flow processes under the action of acoustic emission pulses.

These estimates prove the applicability of the proposed model for the formation of autowaves of localized plastic flow. In this case, we can assume that the autowave equations (3.3) and (3.4) describe the kinetics of events in the dynamic and signal subsystems of the deformable medium, respectively. They clearly indicate the close relationship of plastic and elastic phenomena.

The principal problem arising in explaining the nature of large-scale periodicity in the location of the deformation localization centres is the matching of the autowave scale $\lambda \approx 10$ mm with the scale of dislocation processes of $10^{-6} \leq b \leq 10^{-4}$ mm Within the framework of the model being developed, this explanation boils down to the following. Let the shift emit a transverse elastic wave with a frequency $\omega_m \approx 10^6$ Hz, corresponding to the maximum intensity in the spectrum of the acoustic signal during deformation [Williams, 1980]. It is known [Tokuoka, and Iwashimizu, 1968] that, passing through an elastically intense region, such a wave splits into two orthogonally polarized components propagating at speeds of v_1 and $v_2 \neq v_1$ with lengths of $L_1 = v_1/\omega_m$ and $L_2 = v_2/\omega_m$. The difference in wavelengths in this case

$$\delta L = L_2 - L_1 \approx \frac{v_2 - v_1}{\omega_m} \approx \frac{\sigma_2 - \sigma_1}{2\omega_m \rho V_t} \tag{4.3}$$

is ~10^{-4} m, provided that in equation (4.3) the difference between the main normal stresses is $\sigma_2 - \sigma_1 \approx 10^8$ Pa, the density of the material is $\rho \approx 5\times10^3$ kg/m^3, and the speed of sound is $V_t \approx 3\times10^3$ m/s. The probability of activating a new shift increases when the maxima of the squares of the stresses in both waves coincide, that is, at the highest elastic energy. This corresponds to the condition $L^2/\delta L \approx \lambda \approx 10^{-2}$ m, which is close to the observed autowave length and explains the generation of plasticity nuclei at a distance ~λ from the existing deformation front due to the acoustic induction of deformation.

Another assessment is based on the analysis of the propagation of a acoustic emission pulse (an acoustic signal) through a zone with a nonuniform dislocation density, for example, an existing centre of plastic flow (Fig. 4.5 a). Simplifying, we assume that the dislocation density in each of the fragments decreases from the centre to the periphery. By virtue of the well-known relation $\sigma_i \approx Gb\rho_{disl}^{1/2}$ (Friedel, 1967), internal stresses σ_i are also non-uniformly distributed in such a zone. Such a region in a deformable medium can be viewed as an acoustic lens with a diameter *C*. In combination with the existence of the dependence $V_t \sim \sigma_i$ [Murav'ev, Zuev, Komarov, 1996], this causes

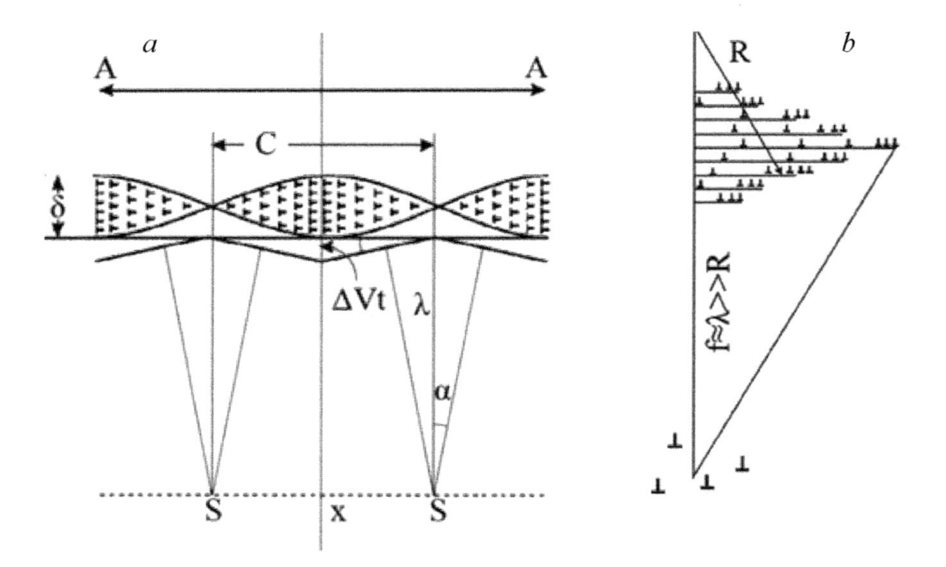

Fig. 4.5. Distribution of inhomogeneities of the dislocation structure, which play the role of acoustic lenses (*a*); acoustic lens diagram (*b*).

the rotation of the front sections of the plane *A–A* wave passing through such an area by a small angle α, as shown in Fig. 4.5 a. In this case, it turns out that the waves from neighbouring regions that play the role of lenses are focused on the axis of symmetry, where the level of elastic stresses increases and, accordingly, the probability of relaxation plasticity acts. This initiates the formation of a new deformation nuclei at a distance of $\sim\lambda$ from the initial one.

A simple geometric calculation, for which the details and designations are explained in Fig. 4.5 a, shows that

$$\lambda = \frac{C}{2\sin\alpha} \approx \frac{C}{2\tan\alpha} \approx \frac{1}{\Delta V_l \delta / CV_t} \approx C \cdot \frac{C}{2\delta} \frac{V}{\Delta V_t} \tag{4.4}$$

It is convenient to make a quantitative estimate for polycrystalline Al, in which the transverse sound velocity is $V_t \approx 3\times10^3$ m/s, and its experimentally found change in the plastic strain interval corresponding to the parabolic hardening stage, $\Delta V_t \leq 10$ m/s [Murav'ev, Zuev, Komarov, 1996]. When the size of the dislocation cell is $p \approx 10^{7}$ m and the ratio $C/2\delta \approx 10$, we obtain $\lambda \approx 10^{-2}$ m, which is close to the experimentally observed distance between the localized deformation sites. Since $\lambda \gg C > b$, we can assume that relation (4.4) connects the microscale of the dislocation substructure and the macroscale of localized deformation regions.

A simpler version of such an assessment is possible. Since the velocity of elastic waves in a medium depends on the deformation [Zuev, Semukhin, 2002], and the dislocations are usually distributed non-uniformly, forming ensembles of different shapes and sizes [Hirsch et al., 1968; Kozlov, Starenchenko, Koneva, 1993], the area of heterogeneity of size l_{disl} can be considered as an acoustic lens with a radius of curvature $R \approx l_{disl}$. Its focal length f, according to, for example, Born and Wolf [1970], will be (Fig. 4.5 b)

$$f = \frac{R}{\upsilon - 1}, \tag{4.5}$$

where $\upsilon = V_0/V$ is the refractive index of sound waves. From the experimental data [Zuev, Semukhin, 2002] it follows that almost to fracture $\upsilon \approx 1.002$; during the deformation of Al $R \approx l_{disl} \approx 10^{-2}$ mm [Myshlyaev, 1972]. Then, according to the formula (4.5), $f \approx 5$ mm. At this distance $f \approx \lambda$, the additional elastic energy is concentrated and the probability of the stress concentrator decay and the occurrence of the plasticity event increases. It is here that a new centre of deformation localization begins to develop. Since the values of υ and R depend on the structure and properties of the material, their evolution during the plastic flow can determine the restructuring of the wave pattern of strain localization. The role of acoustic lenses can be played by any dislocation assemblies with a non-uniform distribution of dislocations (and stresses) over the volume   dislocation tangles, cells, etc.

In these cases, the variants of distribution and behaviour of macroscopic zones of localized deformation are associated with changes in the geometry of acoustic lenses (values of C, δ and their ratio $C/2\delta$) or the distribution of dislocations in them during plastic flow. Thus, in accordance with relations (4.4) and (4.5), an increase in the fragment size initiates an increase in λ and may cause the movement of the source of plastic flow along the stretching axis at the stage of linear strain hardening.

The estimates show that, within the framework of the proposed two-component model for the development of a localized plastic flow, there is a consistent explanation for the emergence of a macroscopic autowave scale in a real deformable material whose defects — dislocations — have a much smaller spatial scale $\sim b$.

4.3. The basic equation of the model – the elastoplastic invariant

To get a formal description of the two-component model, we will try to find a quantitative ratio that corresponds to its structure and allows us to take into account and explain its most significant part – the relationship (interdependence) of elastic and plastic deformation of materials. To this end, we consider the characteristics of the space-time processes that determine the functioning of the model, that is, the autowaves of localized plasticity and elastic waves.

4.3.1. Introduction of an elastoplastic strain invariant

The phase autowaves of plastic flow localization, observed at the stages of linear strain hardening of a number of materials (see Table 3.2), were characterized by experimentally determined length and propagation velocity λ and V_{aw}, respectively [Zuev, Danilov, Barannikova, 2008; Zuev, 2001, 2007, 2012]. A technique for experimentally estimating these values is given in Chapter 2.

Table 4.1. Comparison of χV_t and λV_{aw} (given in $\times 10^7$ m^2/s)

	Metals and alloys											
	Cu	Zn	Al	Zr	Ti	V	Nb	α-Fe	γ-Fe	Ni	Co	Mo
λV_{aw}	3.6	3.7	7.9	3.7	2.5	2.8	1.8	2.55	2.2	2.1	3.0	1.2
χV_t	4.8	11.9	7.5	11.9	7.9	6.2	5.3	4.7	6.5	6.0	6.0	7.4
$\lambda V_{aw}/\chi V_t$	0.75	0.3	1.1	0.3	0.3	0.45	0.33	0.54	0.34	0.35	0.5	0.2

	Metals and alloys							Alkali-halide single crystals			Rocks	
	Sn	Mg	Cd	In	Pb	Ta	Hf	KCl	NaCl	LiF	Marble	Sandstone
λV_{aw}	2.4	9.9	0.9	2.6	3.2	1.1	1.0	3.0	3.1	4.3	1.75	0.6
χV_t	5.3	15.8	3.5	2.2	2.0	4.7	4.2	7.0	7.5	8.8	3.7	1.5
$\lambda V_{aw}/\chi V_t$	0.65	0.63	0.2	1.2	1.6	0.2	0.24	0.43	0.4	0.5	0.5	0.4

The processing of experimental data for the quantities λ and V_{aw} showed that the products $\lambda \cdot V_{aw}$ for different metals are close to each other. This follows directly from the data of the corresponding calculations of these works, given in Table 4.1 for all studied metals and alloys. The average value of the products of these quantities is $\langle \lambda V_{aw} \rangle \approx 10^{-7}$ m^2/s.

These data, originally obtained for the stages of linear strain hardening, were supplemented by the results of similar processing of patterns of localized plasticity observed at the easy slip stage in single crystals of Cu, Fe_{II}, and Sn. In this case, when measuring the quantities λ and V_{aw}, it was also established that $\langle \lambda V_{aw} \rangle_{eg} \approx 10^{-7}$ m^2/s.

The deformation of the single crystal of the TiNi intermetallic compound of equiatomic composition is due to the phase transformation [Otsuka, Shimizu, 1986; Boyko, Garber, Kosevich, 1991]. Experimental estimation of the parameters of the localized plasticity observed in this case also led to $\langle \lambda V_{aw} \rangle_{pt} \approx 10^{-7}$ m^2/s [Zuev, 1996].

When analyzing experimental data on the motion of individual dislocations in single crystals [Nadgorny, 1972; Lubenets, 1973; Caillard, 2010] it turned out that the characteristic lengths of dislocation paths are $10^{-5} \leq l \leq 10^{-4}$ m, and the speed of movement of dislocations is $10^{-3} \leq V_{disl} \leq 10^{-2}$ m/s, so that $\langle l \cdot V_{aw} \rangle_{pt} \approx 10^{-7}$ m^2/s. The constancy of this product is explained by the fact that individual dislocations with a density ρ_m and Burgers vector b begin to move if the applied external stresses reach the level of internal stresses created by the dislocation network [Friedel, 1967; Suzuki, Yoshinaga, Takeuchi, 1989], that is, provided

$$\sigma \geq \frac{Gb}{2\pi}\rho_m^{1/2}, \tag{4.6}$$

where G is the shear modulus. The average path length of dislocations, determined by the size of the mesh of the dislocation grid, in this case is

$$l \approx \rho_m^{-1/2} = \frac{Gb}{2\pi\sigma} \sim \sigma^{-1}. \tag{4.7}$$

Dislocations move between local barriers during a jump, in a quasi-viscous manner with speed [Alshitz, Indenbom, 1975 a, b]

$$V_{disl} \approx (b/B)\cdot\sigma \sim \sigma, \tag{4.8}$$

where B is the coefficient of quasi-viscous dragging of dislocations by the phonon gas. From equations (4.7) and (4.8) it follows that the product

$$l \cdot V_{disl} = \frac{Gb^2}{2\pi B} \approx \text{const} \tag{4.9}$$

does not depend on stress, and the estimate with characteristic values of $G \approx 30$ GPa, $b \approx 2\cdot10^{-10}$ and $B \approx 10^{-4}$ Pa s shows that $\langle l\cdot V_{aw}\rangle_{disl} \approx 10^{-7}$ m^2/s.

Finally, during the deformation by compression of samples from alkali halide crystals and rocks, a linear strain hardening stage was also observed. As follows from Table 4.1, in this case $\langle l\cdot V\rangle_{ahc} \approx 3\cdot10^{-7}$ m^2/s, and $\langle l\cdot V\rangle_{rock} \approx 10^{-7}$ m^2/s.

Obviously, from the listed data follows

$$\langle\lambda V_{aw}\rangle_{lwh} \approx \langle lV\rangle_{disl} \approx \langle\lambda V_{aw}\rangle_{eg} \approx \langle\lambda V_{aw}\rangle_{pt} \approx \langle\lambda V_{aw}\rangle_{ahc} \approx \langle\lambda V_{aw}\rangle_{rock} \approx 10^{-7}\,\text{m}^2/\text{s} \tag{4.10}$$

where the indices have the following meaning: *lwh* – linear work hardening, *disl* – dislocation, *eg* – easy glide, *pt* – phase transformation, *ahc*– alkali-halide crystals, *rock* – rocks. Thus, for elastic deformation, the product of characteristic lengths and speeds at the stages of linear strain hardening and light slip, during deformation due to phase transformation and movement of individual dislocations, as well as for deformation of alkali halide crystals and rock samples, is ~10^{-7} m^2/s.

The characteristics of elastic waves used here were the interplanar distances in the crystal lattice χ, corresponding to the maximum intensities of X-ray reflexes [Mirkin, 1961], and the propagation velocity V_t of transverse elastic waves, given in Anderson [1968]. As one can see from Table 4.1, $\langle \chi V_t \rangle \approx 10^{-7}$m^2/s.

Expression (4.10) can be given a dimensionless form valid for each type of deformation by entering the parameter $\hat{Z}$ in terms of the relations $\frac{\langle \lambda \cdot V_{aw} \rangle_{lwh}}{\langle \chi \cdot V_t \rangle_{el}} = \hat{Z}_{lwh}, \sqrt{a^2+b^2}$, $\frac{\langle l \cdot V \rangle_{disl}}{\langle \chi \cdot V_t \rangle_{el}} = \hat{Z}_{disl}$, $\frac{\langle \lambda \cdot V_{aw} \rangle_{eg}}{\langle \chi \cdot V_t \rangle_{el}} = \hat{Z}_{eg}$,
(4.11)

$\frac{\langle \lambda \cdot V_{aw} \rangle_{pt}}{\langle \chi \cdot V_t \rangle_{el}} = \hat{Z}_{pt}$, $\frac{\langle \lambda \cdot V_{aw} \rangle_{ahk}}{\langle \chi \cdot V_t \rangle_{el}} = \hat{Z}_{ahc}$ and $\frac{\langle \lambda \cdot V_{aw} \rangle_{rock}}{\langle \chi \cdot V_t \rangle_{el}} = \hat{Z}_{rock}$. It was established that $\hat{Z}_{lwh} \approx \hat{Z}_{disl} \approx \hat{Z}_{eg} \approx \hat{Z}_{pt} \approx \hat{Z}_{ahc} \approx \hat{Z}_{rock} \approx 1/2$, and the following general expression was obtained:

$$\left\langle \frac{\lambda \cdot V_{aw}}{\chi \cdot V_t} \right\rangle = \text{const} = \hat{Z} = \frac{1}{2} \pm \frac{1}{4}, \qquad (4.11)$$

which includes all the deformation variants listed above. Expression

(4.11) was called the **elastoplastic deformation invariant** [Zuev, 2011, 2015; Zuev, 2001, 2007]. Graphic representation of the discussed data is shown in Fig. 4.6.

The relation (4.11) indicates the scale invariance of the plastic flow, at least for the three studied scale levels and different deformation mechanisms: linear strain hardening, easy glide, deformation due to the movement of individual dislocations and deformation of the phase transformation. In addition, phase autowaves in single crystals of Fe_{II} were observed upon deformation by twinning [Zuev, Danilov, Baranikova, 2008; Zuev, Barannikova, 2009]. Thus, the elastoplastic invariant acquires the meaning of a universal characteristic of plastic deformation processes.

The physical meaning of the invariant (4.11) is that it relates the characteristics of elastic (χ and V_t) and plastic (λ and V_{aw}) strains. At the same time, equation (4.11) is non-linear, since the magnitude of the speed of sound in it V_t depends in a complex way on strain and stress (Fig. 4.7) [Zuev, Semukhin, 2002; Zuev, Semukhin, Lunev, 2004; Kobayashi, 2010]. This is important for describing the processes of plastic forming.

4.3.2. Elastoplastic invariant and characteristics of the medium

It is interesting to find the connection between invariant (4.11) and other (except χ and V_t) lattice characteristics of deformable media. For this, as a first step, the known relations $G \approx \chi^{-1} \cdot d^2W/du^2$ and $V_t^2 \approx G/\rho \approx \chi^2\omega_D^2$ can be used, where W is the interparticle potential, u is a small displacement, ρ is the density of a deformable medium , and w_D is the Debye frequency [Ashcroft, Mermin, 1979; Kosevich, 1981; Newnham, 2005]. In this case

$$\lambda V_{aw} \approx \hat{Z} \cdot \frac{d^2W/du^2}{(\omega_D\chi)\rho} \approx \hat{Z}\frac{d^2W/du^2}{\xi_1}, \tag{4.12}$$

the quantity $\xi_1 = (\omega_D\chi)\rho = V_t\rho$ appearing in relation (4.12) is the acoustic resistance of the medium [Lüthi, 2007].

In addition, the invariant (4.11), written in the form

$$\frac{\lambda}{V_t} \approx \hat{Z} \cdot \frac{\chi}{V_{aw}} \approx \hat{Z}\vartheta \approx 10^{-5}\,\text{s}, \tag{4.13}$$

indicates the equality of the characteristic times of development of elastic and plastic deformation acts ϑ. Assuming that plastic deformation develops in a thermally activated manner [Caillard, Martin, 2003], we write

$$\vartheta \approx \omega_D^{-1} \exp(U/k_BT). \tag{4.14}$$

If $k_BT \approx$ 1/40 eV, then the activation energy is U $\approx$ 0.5 eV, which is typical of many acts of plastic flow [Friedel, 1967; Caillard, Martin, 2003]

The product λV_{aw} linearly depends on the interplanar distance, as shown in Fig. 4.8. This dependence is split into two, one of which refers to the fourth period, and the second covers the elements of the fifth and sixth periods of the Periodic Table of Elements.

We also note that the quantities V_{aw} and λV_t have the dimension $L^2 \cdot T^{-1}$ corresponding to the kinematic viscosity or diffusion coefficient. The dynamic viscosity estimate

$$\rho\lambda V_{aw} \approx \hat{Z}\rho\chi V_t \approx 10^{-4}\,\text{Pa s} \tag{4.15}$$

Finally, writing the right-hand part of equation (4.11) in the form $\hat{Z} \cdot \chi \cdot V_t = \hat{Z} \cdot \chi^2 \cdot \omega_D$ and using the well-known relationship $k_B\theta_D$ =

$\hbar\omega_D \approx h(V_t/\chi)$, where $\hbar$ is the Planck constant, and θ_D is the Debye parameter (Debye temperature), we obtain the equation

$$\lambda V_{aw} \approx \hat{Z}\cdot\chi V_t \approx \hat{Z}\chi^2\frac{k_B\theta_D}{\hbar} \approx \hat{Z}\frac{k_B}{\hbar}\chi^2\theta_D(T), \quad (4.16)$$

introducing temperature dependence of localization characteristics of plastic flow (λ, V_{aw} or their product $\lambda\cdot V_{aw}$) through temperature dependence $\theta_D(T)$ [Ashcroft, Mermin, 1979]. The relation (4.16) qualitatively correctly describes the experimentally observed temperature variation of autowave characteristics [Zuev, Danilov, Barannikova, 2008].

The generality of the elastoplastic deformation invariant written in the form of equation (4.11) for different deformation mechanisms and its connection with the lattice characteristics of materials makes it possible to give relation (4.11) the meaning of a mathematical formulation of a two-component model for the development of a localized plastic flow.

4.3.3. On the nature of the elastoplastic deformation invariant

In this case, the problem of the physical meaning of the elastic-plastic invariant (4.11) becomes more important. Consider some thoughts about its nature. As mentioned, the localization of plastic deformation is a consequence of the self-organization of a nonlinear active deformable medium containing structural defects. A common feature of self-organization processes in an open thermodynamic system is a decrease in entropy in such a process. As shown in section 3.1.3. [Zuev, 2005], this condition is fulfilled when generating autowaves of the localized plastic flow, so using the entropy factor to determine the physical nature of plastic deformation localization processes is quite reasonable.

As noted in Chapter 1, the plastic flow process is based on competition between the processes of the birth and decay (relaxation) of elastic stress concentrators. This means that during the deformation, the space–time distributions of the elastic stress fields $\sigma(x,y,t)$ and plastic strainss $\varepsilon(x,y,t)$ are transformed in an interdependent manner, and, in accordance with the elastoplastic invariant, the velocities V_t and V_{aw} control the kinetics of the transformation processes corresponding to fields, and the lengths χ and λ define the *spatial scales* of the redistribution processes.

Then the elastoplastic invariant should be presented in the form

$$\frac{\lambda/\chi}{V_t/V_{aw}} = \frac{p_{scale}}{p_{kin}} = \hat{Z}, \tag{4.17}$$

where the ratios $\lambda/\chi = p_{scale} > 1$ and $V_t/V_{aw} = p_{kin} > 1$ can be considered as scale and kinetic thermodynamic probabilities [Kubo, 1970; Klimontovich, 1999, 2002]. The *scale* thermodynamic probability p_{scale} determines the number of possible places for the origin of the autowave of a localized plastic flow in a deformable medium and is related to the difference in the spatial scales of the elastic and plastic deformation processes. In turn, the *kinetic* thermodynamic probability p_{kin} determines the choice by a deformable system of the required autowave speed from the interval of its possible values

In view of what was said, the equation

$$\ln \hat{Z} = \ln p_{scale} - \ln p_{kin}, \tag{4.18}$$

which with the help of the Boltzmann formula can be transformed into equations for entropy changes associated with the scale difference

$$\ln \hat{Z} = \ln p_{scale} - \ln p_{kin}, \tag{4.19}$$

and the speed difference

$$\Delta S_{kin} = k_B \ln \frac{V_t}{V_{aw}} = k_B \ln p_{kin}. \tag{4.20}$$

Finally, from equations (4.18) - (4.20) we have

$$\Delta S = -k_B \ln p_{kin} + k_B \ln p_{scale} = -\Delta S_{kin} + S_{scale} = k_B \ln \frac{1}{2} < 0, \tag{4.21}$$

whence the equivalence of invariant (4.11) follows from the statement that in the process of generating a phase autowave, the entropy of the deformed system changes by $-\Delta S = \Delta S_{scale} - \Delta S_{kin}$ (decreases). The signs of the terms $\Delta S_{scale} > 0$ and $\Delta S_{kin} < 0$ in equation (4.21) determine the difference in the contributions of scale and kinetic factors to the nature of localized plastic deformation. The contribution from the scale differences $\lambda/\chi = p_{scale}$ is dissipative, because it is equivalent to the presence of the structure of the medium, and this factor is the cause of the appearance of dispersion and dissipative processes in general [Trubetskov, 2003]. On the contrary, the contribution from the speed difference $V_t/V_{aw} = p_{kin}$, which reduces the total entropy of the system, apparently contributes to the self-organization of the medium.

The value $\Delta S < 0$ characterizes the overall decrease in entropy during the formation of the phase autowave and the localization of the plastic flow. Since

$$\hat{Z} = \exp\left(\Delta S/k_B\right) \approx \frac{1}{2}, \tag{4.22}$$

then $\Delta S = k_B \cdot \ln 1/2 \approx 0.7\ k_B$ on the relaxation act [Slutsker, 2005, 2006; Gilyarov, Slutsker, 2010a, b].

We now consider the evolution of the fields of elastic stresses and plastic strains mentioned above, and we formulate ideas about the physical nature of the invariant (4.11) based on thermodynamic considerations. To do this, we assume that the displacement rates for space–time transformations of the strain fields and stresses in a deformable system with small deviations from equilibrium are linear in the gradients of plastic and elastic strains, that is, $\dot{u}_{pl}^{(p)} \approx D_{\varepsilon\varepsilon}\nabla\varepsilon_{pl}$ and $\dot{u}_{el}^{(p)} \approx D_{\sigma\sigma}\nabla\varepsilon_{el}$ respectively [Nye, 1960; Rumer, Ryvkin, 2000]. To simplify the recording, by virtue of Hooke's law, $\sigma = E\varepsilon_{el} \sim \varepsilon_{el}$ instead of elastic stresses we use the elastic strains. In addition, due to the connection of strain and stress, given by the relationship $\sigma(\varepsilon)$, additional flows $\dot{u}_{el}^{(ad)} \approx D_{\varepsilon\sigma}\nabla\varepsilon_{pl}$ and $\dot{u}_{pl}^{(ad)} \approx D_{\sigma\varepsilon}\nabla\varepsilon_{el}$ [De Groot, Mazur, 1964]. This makes it possible to create a system of equations for the plastic and elastic component of the displacement velocities.

$$\dot{u}_{pl} = D_{\varepsilon\varepsilon}\nabla\varepsilon + D_{\varepsilon\sigma}\nabla\varepsilon_{el}, \tag{4.23}$$

$$\dot{u}_{el} = D_{\sigma\varepsilon}\nabla\varepsilon_{el} + D_{\sigma\sigma}\nabla\varepsilon. \tag{4.24}$$

The coefficients of eqs. (4.23) and (4.24) form a matrix $\begin{pmatrix} D_{\varepsilon\varepsilon} & D_{\varepsilon\sigma} \\ D_{\sigma\varepsilon} & D_{\sigma\sigma} \end{pmatrix}$

According to the principle of symmetry of the Onzager kinetic coefficients [Nye, 1960; Rumer, Ryvkin, 2000; Landau, Lifshits, 2002], its nondiagonal components are equal, that is, $D_{\varepsilon\sigma} = D_{\sigma\varepsilon}$. In this case, the diagonal coefficients of the matrix $D_{\varepsilon\varepsilon}$ and $D_{\sigma\sigma}$ do not have to be equal: from equations (3.17) and (3.18) it follows that

$D_{\varepsilon\varepsilon} \ll D_{\sigma\sigma}$. The coefficients $D_{\varepsilon\sigma}$ and $D_{\varepsilon\varepsilon}$ and $D_{\sigma\varepsilon}$ and the products λV_{aw} and λV_t in the invariant (4.11) have the dimension L^2 T^{-1}. If we assume that $\lambda\ V_{aw} \equiv D_{\varepsilon}$ and $\chi\ V_t \equiv D_{\sigma}$, then the invariant (4.11) takes the form

$$\frac{D_{\varepsilon\sigma}}{D_{\sigma\varepsilon}} = \hat{Z}, \tag{4.25}$$

reducing the elastoplastic deformation invariant to the ratio of the off-diagonal coefficients of the matrix of the system of equations (4.23) and (4.24).

Thus, we can assume that the meaning of the elastoplastic deformation invariant (4.11), which relates the propagation characteristics of elastic waves and autowaves of a localized plastic flow, is determined by the entropy factor.

a

4.4. Implications of the two-component model

The above considerations about the structure of the two-component model, its mathematical formulation in the form of equation (4.11), and data on the nature of the elastoplastic invariant deformation indicate their fundamental importance for understanding the nature of plastic deformation. In particular, processing the equation of the elastoplastic invariant (4.11) shows that explanations of many experimentally observed patterns of development of a localized plastic flow follow from it [Zuev, Danilov, Semukhin, 2002; Zuev, Danilov, 2003; Zuev, Danilov, Barannikova, 2008; Danilov, Zuev, 2008; Zuev, 2015]. Let us consider successively a series of consequences from the two-component model of localized plasticity and from the elastoplastic strain invariant equation (4.11).

4.4.1. Phase autowave propagation speed

Differentiating equation (4.11) by deformation ε, we get

$$\lambda \frac{dV_{aw}}{d\varepsilon} + V_{aw}\frac{d\lambda}{d\varepsilon} = \hat{Z}\cdot\chi\frac{dV_t}{d\varepsilon} + \hat{Z}\cdot V_t\frac{d\chi}{d\varepsilon} \tag{4.26}$$

Writing this equation for V_{aw}, we have

$$V_{aw} = \left(\frac{d\lambda}{d\varepsilon}\right)^{-1}\left(\hat{Z}\cdot\chi\frac{dV_t}{d\varepsilon} + \hat{Z}\cdot V_t\frac{d\chi}{d\varepsilon} - \lambda\frac{dV_{aw}}{d\varepsilon}\right) \tag{4.27}$$

Since $\chi = \text{const}$, we have $\dot{Z} V_t\, d\chi/d\varepsilon = 0$. Then

$$V_{aw} = \hat{Z}\chi \cdot \frac{dV_t}{d\lambda} - \lambda \cdot \frac{dV_{aw}}{d\lambda}. \tag{4.28}$$

By simple transformations, equation (4.28) reduces to

$$V_{aw} = \hat{Z}\chi \cdot \frac{dV_t}{d\lambda} - \chi \cdot \frac{dV_{aw}}{d\lambda} \cdot \frac{\lambda}{\chi} \approx V_0 + \frac{\Xi}{\theta}, \tag{4.29}$$

where the scale thermodynamic probability introduced above, written as $\lambda/\chi)^{-1}$, in accordance with the work of Roitburd [1972], is identified with the strain hardening coefficient, that is, $\lambda/\chi)^{-1}$–θ<<1. In the earlier theory of hardening, developed by Seeger [1960], the strain hardening coefficient at the linear hardening stage is also considered as a ratio of small and large scales, but for this case $\theta \approx \sqrt{\chi/\lambda} \approx 10^{-3}$, which is typical for this stage of the process. Taking these considerations into account, equation (4.29) leads to V_{aw} = V_0+Ξ/0 as it was established experimentally earlier [Zuev, Danilov, Barannikova, 2008].

4.4.2. Dispersion of phase localized deformation autowaves

We write equation (4.11) in the form

$$V_{aw} = \frac{\Theta}{\lambda} = \frac{\Theta}{2\pi} \cdot k \tag{4.30}$$

where $\Theta = \hat{Z}\chi V_t$. If $V_{aw} = d\lambda/dT = d\omega/dk$ then $d\omega = (\Theta/2\pi)\, k{\cdot}dk$. In this case

$$\int_{\omega_0}^{\omega} d\omega = \frac{\Theta}{2\pi} \int_{0}^{k-k_0} k \cdot dk, \tag{4.31}$$

and, if we assume that $\Theta/4\pi \equiv \alpha$, then the dispersion law of autowaves of a localized plastic flow acquires a quadratic form

$$\omega(k) = \omega_0 + \frac{\Theta}{4\pi}(k - k_0)^2, \tag{4.32}$$

the existence of which was previously established experimentally by Barannikova [2004], for autowaves of localization of the plastic flow in iron single crystals and aluminum polycrystals.

4.4.3. Constants in the dispersion relation for autowaves

Giving equation (4.16) the form

$$V_{aw} \approx \chi^2 \frac{k_B \theta_D}{\hbar} \cdot \frac{1}{\lambda} \approx \chi^2 \frac{k_B \theta_D}{\hbar} \cdot k \approx \varsigma \cdot k \tag{4.33}$$

We can calculate $\varsigma = \chi^2 \cdot k_B \theta_D / \hbar \approx \chi^2 \cdot \omega_D$ using the value χ and the values of the Debye parameter $\theta_D^{(Fe)} = 420$ K and $\theta_D^{(Al)} = 394$ K [Ashcroft, Mermin, 1979], and $\zeta^{(Fe)} \approx 3.7 \cdot 10^{-7}$ m²/s, and $\zeta^{(Al)} \approx 4.45 \cdot 10^{-7}$ m²/s,. This is in satisfactory agreement with the values (1.0 ± 0.08) · 10–7 m²/s (12.9 ± 0.15) · 10^{-7} m²/s calculated from experimental data on the dependence given in the book by Zuev, Danilov, Barannikova [2008].

4.4.4. Connection of the autowave length with the grain size in a polycrystal

We write equation (4.11) in the form

$$\lambda = \hat{Z}\chi \cdot \frac{V_t}{V_{aw}} \tag{4.34}$$

and take into account that the speeds V_t and V_{aw} depend on the grain size. In this case, the differentiation of the relation (4.34) by δ gives

$$\frac{d\lambda}{d\delta} = \hat{Z}\chi \cdot \frac{d}{d\delta}\left(\frac{V_t}{V_{aw}}\right) = \hat{Z}\chi \cdot \left(\frac{V_{aw} \cdot dV_t/d\delta - V_t \cdot dV_{aw}/d\delta}{V_{aw}^2}\right) \tag{4.35}$$

From equation (4.35) it follows

$$d\lambda = \hat{Z}\chi \cdot \left(\frac{dV_t}{d\delta} \cdot \frac{1}{V_{aw}} - V_t \frac{dV_{aw}}{d\delta} \cdot \frac{1}{V_{aw}^2}\right) \cdot d\delta = \left(a_1\lambda - a_2\lambda^2\right) \cdot d\delta, \tag{4.36}$$

where $a_1 = \frac{1}{V_t} \cdot \frac{dV_t}{d\delta} = \frac{d \ln V_t}{d\delta}$ and $a_2 = \frac{1}{\hat{Z}\chi V_t} \cdot \frac{dV_{aw}}{d\delta}$ since $V_{aw} = \hat{Z}\chi V_t \cdot \frac{1}{\lambda}$.
The solution of the differential equation (4.36) is a logistic function or a Verhulst function [Volterra, 1976]

Table 4.2. Checking the fulfillment of the invariant relation (4.11) for two ranges of grain sizes in aluminium

Grain size range	$\chi V_t \cdot 10^7$	$\lambda V_{disl} \cdot 10^7$	$\lambda V_{disl}/\chi V$
mm	m²/s		
$0.05 \le \delta \le 0.1$	5.1	2.6	~1/2
$0.1 \le \delta \le 5$	6.1	3.1	~1/2

$$\lambda(\delta) = \lambda_0 + \frac{a_1/a_2}{1 + C\exp(-a_1\delta)}, \tag{4.37}$$

where λ_0 = const, and C is the integration constant. The dependence (4.37) was experimentally obtained for aluminum [Zuev, Zarikovskaya, Fedosova, 2010] for a grain size range of $5 \cdot 10^{-3} \leq \delta \leq 15$ mm. Its form corresponding to equation (4.37) is shown in Fig. 3.14.

4.4.5. Scale effect for autowave localized plasticity

Differentiate the expression for the invariant (4.11) along the length of the samples L and we obtain the equality

$$\frac{d}{dL}(\lambda V_{aw}) = \frac{d\lambda}{dL}V_{aw} + \lambda\frac{dV_{aw}}{dL} = \hat{Z}\frac{d}{dL}(\chi V_t) = 0, \tag{4.38}$$

which is true, since, obviously, the interplanar distance χ and the speed of sound V_t do not depend on L. In this case

$$\frac{d\lambda}{dL} = -\frac{\lambda}{V_{aw}}\frac{dV_{aw}}{dL}. \tag{4.39}$$

When $\lambda = \lambda_0$ = const and $\frac{dV_{aw}}{dL} \approx \frac{V_{aw}}{L}$ we arrive at $\frac{d\lambda}{dL} \approx \frac{\lambda_0}{V_{aw}} \cdot \frac{V_{aw}}{L}$ and $d\lambda \sim dL/L$, that is, to the experimentally established dependence $\lambda \sim \ln L$ [Zuev et al., 2010].

4.4.6. Autowave equation of localized plasticity

We write the invariant (4.11) in the form

$$\chi/\lambda = \hat{Z}(V_{aw}/V_t) \tag{4.40}$$

and assume that $\varepsilon \approx \lambda/\chi >> 1$ is the plastic strain. Applying the operator $\partial/\partial t = D \cdot \partial^2/\partial x^2$ respectively, we obtain

$$\frac{\partial\varepsilon}{\partial t} = \hat{Z}D\left(-V_t \cdot \frac{\partial^2 V_{aw}^{-1}}{\partial x^2} + V_{aw}^{-1} \cdot \frac{\partial^2 V_t}{\partial x^2}\right) = \hat{Z}D\left[-V_t \cdot \frac{\partial^2 V_{aw}^{-1}}{\partial x^2} + \frac{\partial^2 (V_t/V_{aw})}{\partial x^2}\right]. \tag{4.41}$$

Note that the operator $\partial/\partial t = D \cdot \partial^2/\partial x^2$ in an obvious way can be obtained from the Fourier equation $\frac{\partial}{\partial t}y = D\frac{\partial^2}{\partial x^2}y$ [Ango, 1967]. If we assume that the propagation velocity of ultrasound is $V \approx$ const, and $V_t/V_{aw} \approx \hat{Z}^{-1} \cdot \lambda/\chi \approx \varepsilon$ then

$$V_t/V_{aw} \approx \hat{Z}^{-1} \cdot \lambda/\chi \approx \varepsilon \tag{4.42}$$

This ratio is obviously equivalent to the differential reaction-diffusion equation for the strain rate

$$\dot{\varepsilon} = f(\varepsilon,\sigma) + D_{\varepsilon\varepsilon}\varepsilon'', \tag{4.43}$$

which was received and discussed earlier (see section 3.2.2).

4.4.7. Autowaves and the Taylor – Orowan dislocation kinetics equation

The most important problem of the developed autowave approach to plastic flow is to find and discuss its connection with the theory of dislocations [Van Buren, 1962; Friedel, 1967; Hirth, Lothe, 1972]. It is known that dislocation theories of plasticity are based mainly on the use of the Taylor–Orowan equation (1.1), which determines the rate of plastic deformation.

In section 3.2.3. it was shown that equation (1.1) can be considered as a special case of the autowave equation $\dot{\varepsilon} = f(\varepsilon) + D_{\varepsilon\varepsilon}\varepsilon''$.. The same result can be obtained from the equation of the elastoplastic invariant (4.11) or (4.42) as follows. Comparing equations (1.1) and (4.42), we can conclude that the first term on the right-hand side of equation (4.42) $-\hat{Z}DV_t\dfrac{\partial^2 V_{aw}^{-1}}{\partial x^2}$ is similar to the term $b\rho_{\text{in}}V_{disl}$ in the Taylor–Oroan equation, if we assume that $Vt \approx \chi\omega_D \approx b\omega_D$, and $D_{\varepsilon\sigma} = \hat{Z}D_{\sigma\varepsilon} = \hat{Z}\chi V_t$. Let also $\dfrac{\partial^2 V_{aw}^{-1}}{\partial x^2} \approx \dfrac{V_{aw}^{-1}}{x^2}$. With a chaotic distribution of dislocations, $x^{-2} \approx l^{-2} \approx \rho_m$, where l is the dislocation path. Finally

$$-\hat{Z}DV_t\frac{\partial^2 V_{aw}^{-1}}{\partial x^2} \approx -\hat{Z}^2 \cdot \chi V_t^2 \cdot b\omega_D \cdot \frac{V_{aw}^{-1}}{x^2} \approx -\hat{Z}^2 bV_t\frac{V_t/V_{aw}}{l^2} \approx -\hat{Z}^2 V^{*-1} b\rho_{mob}V_t. \tag{4.44}$$

The velocities of dislocations V_d and transverse ultrasonic waves V_t are usually proportional to each other [Nadgorny, 1972; Suzuki, Yoshinaga, Takeuchi, 1989], so suppose that, $V_t \approx \alpha V_d$ where α = const < 1.Then

Table 4.3. Parameters of autowave processes of strain localization

Material	Structural state	$V_{aw} \cdot 10^5$, m/s	$D \cdot 10^7$ m/s	R · 10^9, m
Fe–0.1%C–2% Mn	polycrystal	4.5	8.1	8.3
Cu–10%Ni–6%Sn	single crystal	6.5	7.6	5.2
Ni (equiatomic)	single crystal	1.0	0.8	0.6
Ni_3Mn (ordered)	polycrystal	10.0	13.5	6.8
Fe_I	single crystal	3.5	7.8	4.3
Fe_I–0.5% N	single crystal	2.7	2.0	1.0

$$\frac{\partial \varepsilon}{\partial t} = -\alpha \frac{\hat{Z}^2}{V^*} \cdot b\rho_{mob} V_d + D_\varepsilon \, \partial^2 \varepsilon / \partial x^2 = \alpha b \rho_{mob} V_{disl} + D_{\varepsilon\varepsilon} \, \partial^2 \varepsilon / \partial x^2 . \quad (4.45)$$

Obviously, the Taylor–Orowan equation differs from equation (4.45) in the absence of the $D_{\varepsilon\varepsilon}\ \partial^2\varepsilon/\partial x^2$ term on the right-hand side and should be considered as a special case of the more general equation of dislocation kinetics (4.45), including *hydrodynamic* $f(\varepsilon,\sigma) = b\rho_m V_{disl} \sim V_{disl}$, and diffusion-like $D_{\varepsilon\varepsilon}\, \partial^2\varepsilon/\partial x^2 \sim \partial^2\varepsilon/\partial x^2$ deformation flow components. Accordingly, the dislocation deformation mechanisms based on equation (1.1) are only limiting cases of the autowave model of plastic flow corresponding to low dislocation densities. Only in this limit does the equation (1.1) allows getting the correct results. However, for large plastic strains and high defect densities corresponding to them, it is necessary to apply equation (4.45).

The connection of the autowave equation with the Taylor–Orowan dislocation kinetics equation was established earlier by us [Zuev, Danilov, 1999] when analyzing the structure of the autowave equation (3.3), carried out in Section 3.2.3. Both variants of the derivation of the relation connecting the dislocation and autowave approaches to the problem of plastic deformation confirm the complementary nature of these approaches.

At the same time, the relation (4.45) shows that the dislocation mechanisms of deformation and hardening [Friedel, 1967; Hirth, Lothe, 1972] describe only the *N*-shaped form of nonlinear functions $f(\varepsilon)$ and $g(\sigma)$ in equations (3.3) and (3.4). These functions have a relaxation nature [Malygin, 1999] and describe elementary acts of plastic flow. For this reason, difficulties arise when it is necessary to use the equations for nonlinear functions $f(\varepsilon)$ and $g(\sigma)$ to describe the whole plastic flow over the sample. This complexity can be overcome with the help of equation (4.45).

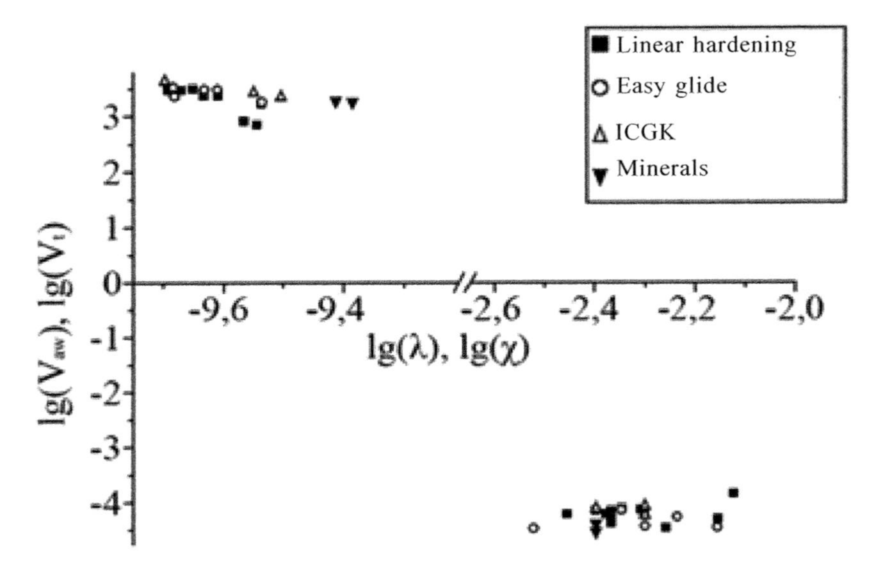

Fig. 4.6. Comparison of the products of the elastoplastic invariant with different laws of deformation.

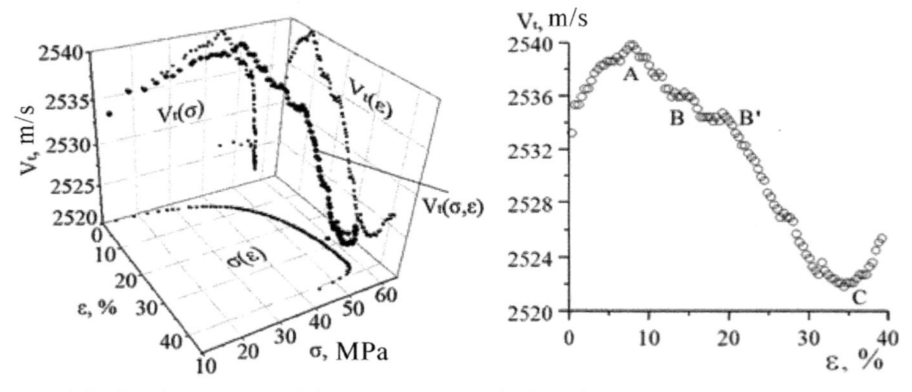

Fig. 4.7. The dependence of the propagation velocity of ultrasound in Al on the strain and stress (a). The speed of sound as a function of strain (*b*); point A corresponds to the limit of proportionality; point *C* – strains with ultimate strength; Section *B–B′* corresponds to the linear stage of strain hardening.

4.4.8. The reason for generating autowaves

Within the framework of the model under discussion, the nature of autowave modes of localized plastic deformation can be explained. In accordance with the Taylor–Orowan equation (1.1), $\dot{\varepsilon} = \alpha b \rho_m V_{disl}$, the condition of the constancy of the strain rate $\dot{\varepsilon}$ = const given by the testing machine is satisfied under the condition $\rho_m V_{disl}$, = const. For the latter, a sufficient density of mobile dislocations ρ_m and a sufficient speed of their movement V_{disl} are necessary.

These conditions can be violated due to a decrease in the density of mobile dislocations [Gilman, 1965] or due to a decrease in their velocity with decreasing effective stress acting on the dislocation from σ to $\sigma - Gb\sqrt{\rho_d}$ during strain hardening with increasing strain [Friedel, 1967; Hirth, Lothe, 1972]. In this case, the condition $\dot{\varepsilon}$ = const can be satisfied only by an additional contribution to the velocity value created by the diffusion-like term $D_{\varepsilon\varepsilon}\varepsilon''$ in equation (4.45), which can be considered as a generalization of the Taylor–Orowan equation. Such a contribution is associated with the generation of new regions of localized plastic flow at a distance of ~λ from the original distance ('throwing' of deformation), which forms the autowave of a localized plastic flow, as shown in Fig. 4.9. This situation was discussed in detail above in section 4.1.3.

4.4.9. Evaluation of linear strain hardening coefficient

We use the relations obtained above for the propagation velocity of the autowave (3.17) and its dispersion (3.22) and write the equality

$$\theta = \frac{\Xi}{2\pi\alpha}\lambda = \frac{\Xi}{2\pi\hat{Z}\chi V_t}\lambda \approx \pi^{-1}\frac{\Xi}{V_t}\cdot\frac{\lambda}{\chi}, \tag{4.46}$$

a calculation by which gives $\theta \approx 3 \cdot 10^{-3}$, which is close to the values observed at the stage of linear strain hardening [Seeger, 1960].

4.4.10. Elastoplastic invariant and Hall–Petch relation

It was previously established that the length of the autowave of localized deformation is related to the grain size [Zuev, Zarikovskaya, Fedosova, 2010]. The mechanical characteristics of metals and alloys – the yield strength, ultimate strength (tensile strength), and flow stress – also significantly depend on this value. To analyze the implementation of the invariant (4.11) in different ranges of grain sizes in aluminium, we must have data on the propagation velocity of

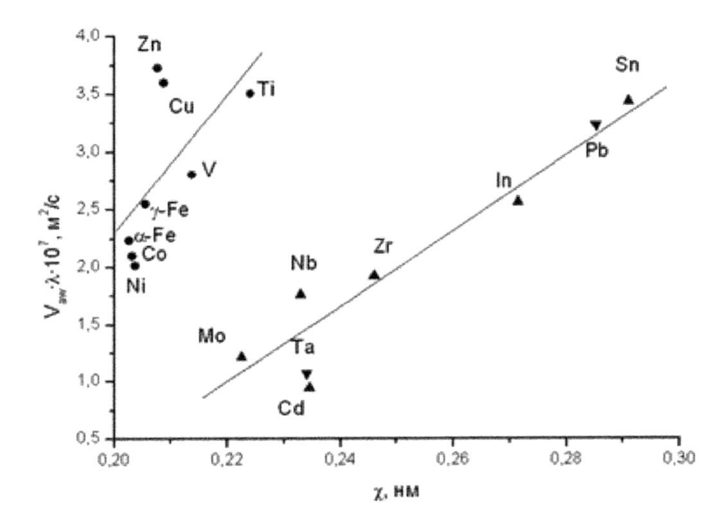

Fig. 4.8. The dependence of the product $V_{aw}\lambda$ on the interplanar distance χ for metals of different periods of the Periodic system of elements: ● – 4th period, ▲ – 5th period, ▼ – 6th period.

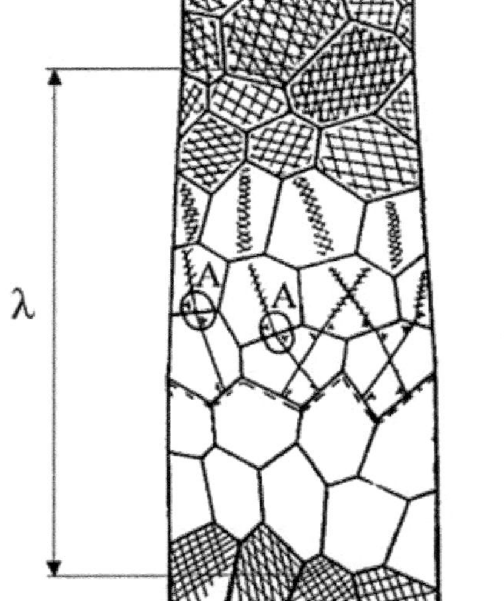

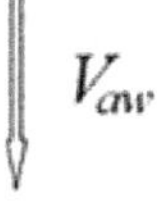

Fig. 4.9. The scheme of "throwing" the strain at a distance ~*l*.

ultrasound in such materials. They were obtained earlier in [Zuev et al., 2013] and are presented in Fig. 3.19, from which it follows that when passing through the boundary grain size of $\delta \approx 0.1$ mm, the propagation velocity of ultrasound drops noticeably. The causes of the fall are associated with the scattering of ultrasound at the grain boundaries. The results of calculations of λV_{aw} values for two ranges of grain sizes are given in Table 4.2. They indicate the validity of the invariant relation in such conditions.

Thus, the fulfillment of the Hall–Petch relation is associated with the automatic adjustment of the acoustic properties of the medium to a change in its structural state. This ensures the implementation of the Hall–Petch relation (3.37) during deformation.

4.4.11. Connection of elastic and plastic component of deformation

Interpreting the physical meaning of the invariant (4.11), we pay attention to the fact that when analyzing the relationship between the elastic ε_{el} and plastic ε_{pl} components of the total strain ε_{tot}, an almost obvious condition is taken $\varepsilon_{tot} = \varepsilon_{el} + \varepsilon_{pl} \approx \varepsilon_{pl}$, since $\varepsilon_{pl} \gg \varepsilon_{el}$. At the same time, the invariant shows that the relationship between the elastic and plastic components of the total strain is not reduced to the generally accepted additive formula, but has a much deeper meaning. According to such ideas, elastic deformation acts as a factor controlling the development of plastic flow. In particular, the formation of the macroscopic scale of localization of plastic deformation in deformable objects is explained by the behaviour of the elastic strain field.

4.4.12. On the relationship of dislocation and mesoscopic scales.

Interpretation of the coefficients $D_{\varepsilon\varepsilon}$ and $D_{\sigma\sigma}$ in the equations (3.3) and (3.4) can be based on the fact that the coefficient $D_{\varepsilon\varepsilon}$ is related to the density of mobile dislocations, and the coefficient $D_{\sigma\sigma}$ is determined by the distribution of stresses. From considerations of dimension for these coefficients can be written

$$D_{\sigma\sigma} \approx \sqrt{\frac{F}{\rho}}, \tag{4.47}$$

where F is the tensile force of the specimen during the test, and

$$D_{\varepsilon\varepsilon} \approx \frac{d}{dt}\rho_m^{-1}. \tag{4.48}$$

The density of mobile dislocations ρ_m, as Gilman showed [Gilman, 1965; Gilman, 1972], is extremely dependent on deformation, so that its time derivative (4.48) can change its sign during deformation.

From equation (4.47) it follows that $D_{\sigma\sigma} \approx 1$ m²/s. The value of $D_{\sigma\sigma}$ according to equation (4.48) is determined less accurately, since the data available in the literature on the density of mobile dislocations at different stages of the process of plastic flow differ significantly. However, based on the available experimental data [Nabarro, Bazinsky, Holt, 1967; Dotsenko, Landau, Pustovalov, 1987], it can be assumed that $10^{-8} \leq D_{\varepsilon\varepsilon} \leq 10^{-7}$ m²/s, so that $D_{\varepsilon\varepsilon} \ll D_{\sigma\sigma}$. Note that the fulfillment of this condition, corresponding to a slower spread of the activator (plastic deformation) as compared to the inhibitor (stress or elastic deformation), is necessary for the realization of the autowave generation process in active media [Vasilyev, Romanovsky, Yakhno, 1989].

As was said, the coefficient $D_{\sigma\sigma}$ describes the redistribution of stresses in the sample volume associated with random walks, and the coefficient $D_{\varepsilon\varepsilon}$ is determined by the rearrangement of the dislocation substructure. In this case, it is natural to consider $D_{\sigma\sigma}$ as a characteristic of the macrolevel of plastic flow, and $D_{\varepsilon\varepsilon}$ is attributed to the kinetics of the development of deformation at the mesolevel. Using the diffusion approximation, we present the coefficients $D_{\varepsilon\varepsilon}$ and $D_{\sigma\sigma} \neq D_{\varepsilon\varepsilon}$ in general form as the product of the path length R and the velocity V, that is, D = RV. Here R is the size of the inhomogeneity region in the deformable system, and V is the rate of redistribution of strains or stresses in it. Since the coefficient $D_{\sigma\sigma}$ is related to the redistribution of stresses, the characteristic speed of this process is the speed of transverse sound V_t, that is, V = $V_t \approx 10^3$ m/s. In this case, R = $l_\sigma \approx D_{\varepsilon\varepsilon}/V_t \approx$ 10–3 m, which can be identified with the mesoscopic scale of plastic strain inhomogeneity.

For the dislocation coefficient $D_{\varepsilon\varepsilon} \approx 10^{-8}$ m²/s, we can put $V \approx V_{disl} \approx 10$ m/s [Lubenets, 1973]. In this case, R = $l_\varepsilon \approx D_{\varepsilon\varepsilon}/V_{disl} \approx 10^{-9}$ m $\approx nb$, where $n \approx 2...5$, which obviously corresponds to the microscopic (dislocation) scale of the plastic flow.

This analysis establishes a hierarchy of structural levels of plastic deformation [Zuev, 1996]. It consists in the fact that the coefficients $D_{\varepsilon\varepsilon}$ and $D_{\sigma\sigma}$, which characterize the macro- and mesoscale of wave structures, respectively, contain lengths that characterize the underlying structural level of plastic flow. The following relations arise that determine the hierarchy and the subordination of the scale of plastic flow:

$$D_{\varepsilon\varepsilon} \approx l_{micro} V_{disl} \text{ – mesolevel} \quad (4.49)$$

$$D_{\sigma\sigma} \approx l_{meso} V_t \text{ – macrolevel} \tag{4.50}$$

These ratios directly link the scale levels of the developing plastic deformation in such a way that the transport factor includes the scale of the underlying level as a multiplier. Thus, the spatial scale of the underlying level determines the kinetics of the processes on the autowave equations (3.3) and (3.4) that lie above the corresponding diffusion coefficient. The results of the calculations are presented in Table. 4.3. In the case of the TiNi alloy, the measured value of V_t was used [Lapshin et al., 1995], since its deformation is due to the martensitic transformation [Otsuka, Shimizu, 1986; Boyko, Garber, Kosevich, 1991], when the speed of sound decreases.

For calculations of the values included in Table 4.3, the coefficient $D_{\varepsilon\varepsilon}$ was estimated from experimental data on the displacement of the deformation fronts δx during time t, that is as $D_{\varepsilon\varepsilon} \approx (\delta x)^2/t$. The thus obtained values of $D_{\varepsilon\varepsilon}$ are used to calculate the scale factor R.

When discussing the ratio of the scales, we also take into account that the autowave in active media is characterized by the presence of a minimum system size l_{min} allowing for the implementation of such a process [Nicolis, Prigogine, 1979]. From the equality of the oscillation period in the system $\vartheta_O \approx 2\pi\omega^{-1}$ and the characteristic diffusion time $\vartheta_D \approx l^2_{min}/2D$ it follows

$$l_{min} \approx (2D\vartheta)^{1/2} \approx (4\pi \cdot D/\omega)^{1/2} \tag{4.51}$$

where D is the transport coefficient. Using for the calculation of l_{min} the characteristics of the slowest processes $D \equiv D_{\varepsilon\varepsilon} \approx 10^{-8}$ m^{2}s and ω 10^{-3} Hz, it is possible to obtain a lower estimate $l_{min} \approx 1.1\times10^{-2}$ m, which is close to the experimentally observed sample length $l^{(exp)}_{min} \leq 2\times10^{-2}$ m, at which it is not possible to register wave processes of strain localization).

4.4.13. Density of mobile dislocations

We now consider the diagonal terms $D_{\varepsilon\varepsilon}$ and $D_{\sigma\sigma} >> D_{\varepsilon\varepsilon}$ of the transport coefficient matrix in the system of equations (4.23) and (4.24). Again, we use relations (4.47) and (4.48). Taking into account that the ratio $D_{\varepsilon\varepsilon}/D_{\sigma\sigma} \approx V_{aw}/V_t \approx 10^{-10}$, we write

$$\frac{d}{dt}\left(\rho_m^{-1}\right) \approx \frac{V_{aw}}{V_t} D_{\sigma\sigma} \tag{4.52}$$

or

$$\frac{d}{dt}\left(\rho_m^{-1}\right) \approx -\rho_m^{-2}\frac{d\rho_m}{dt} \approx \frac{V_{aw}}{V_t} D_{\sigma\sigma}. \tag{4.53}$$

this implies

$$\rho_m^{-2} d\rho_m \approx -D_{\sigma\sigma}\frac{V_{aw}}{V_t} dt \approx -\sqrt{\frac{F}{\rho}} \cdot \frac{V_{aw}}{V_t} \cdot dt$$

or after integration

$$-\rho_m^{-1} = -\sqrt{\frac{F}{\rho}} \cdot \frac{V_{aw}}{V_t} t. \tag{4.55}$$

In this case

$$\rho_m = \sqrt{\frac{\rho}{F}} \cdot \frac{V_t}{V_{aw}} t^{-1}, \tag{4.56}$$

that is, $\rho_m \sim -t^{-1}$. Since with active loading $(V_{mach}/L) \cdot t$, then

$$\rho_m \approx \frac{V_{mach}}{L}\sqrt{\frac{\rho}{F}} \cdot \frac{V_t}{V_{aw}} \cdot \varepsilon^{-1} \approx \frac{V_t}{mL}\sqrt{\frac{\rho}{F}} \cdot \varepsilon^{-1} \sim \varepsilon^{-1}. \tag{4.57}$$

The decrease in the density of mobile dislocations during plastic deformation in the region of large strains according to such a law was described earlier by Gilman [Gilman, 1965; Gilman, 1972].

4.5. Generalization of the two-component plasticity model

The basis of the two-component model of the development of localized plasticity described in this chapter is the idea of the initiation of motion and development of lattice defects by elastic pulses arising as a result of acoustic emission. This model can be given a greater degree of generality for use, including to clarify its relationship with other plasticity mechanisms. This is especially interesting and important when comparing the model with the dislocation mechanisms of plasticity, which can have two aspects.

The first of these is that many researchers have observed the emergence of dislocation microloops near the stress concentrators, capable of developing by two mechanisms: either as sources of slip or as micro-nuclei of destruction [Zuev, 1998]. This view is confirmed by a number of experimental evidence. Thus, for example, Higashida et al. [Higashida et al., 1997, 2004, 2008] observed the birth of dislocations near the crack end due to the work of Frank–Read sources. The occurrence of dislocation defects and microcracks in front of a main crack was experimentally investigated by Finkel [1970] and theoretically described by Raiser [1970].

For twinning, which is one of the main mechanisms for the deformation of materials [Klassen-Neklyudova, 1960; Ishii, Li, Ogata, 2016], it was shown that before the end of the growing twin, prismatic dislocation loops also appear: Mahajan, Green and Brazen [Mahajan, Green, Brasen, 1977] and Mahajan [Mahajan, 1981]. Therefore, there is a reason to believe that microscopic dislocation loops arise in the concentration regions of elastic stresses and are capable of developing when creating suitable conditions. It follows from the developed model that such conditions can be created under the action of acoustic pulses.

Acting on these loops, acoustic impulses initiate the development of deformation. Consider this elementary mechanism in accordance with the concepts of the physical dislocation core developed by Bengus [1966], who showed that, near the dislocation line, a region of $\sim 2b$ in which the material is in a state unstable with respect to the shift (physical dislocation core). In such a region, due to thermal dislocations, paired bends and virtual loops with a radius of less critical $\sim 10^4\ b$ are spontaneously generated. In the absence of external influence of any nature, such a “dislocation - virtual loop” system is static, since the nucleation of a loop in the slip plane is equally likely on both sides of the dislocation. However, the external effect of a non-uniform elastic field of an acoustic impulse can change the equal distribution of statistical loops, which leads to the generation of kinks on one side of the dislocation line, that is, to the beginning of its movement and the development of plastic deformation.

In principle, this mechanism is in agreement with the concepts of the “phonon wind” developed by Alshitz and Indenbom [1975a, b] in explaining the nature of the dislocation retardation during high-speed over-barrier motion. It can also be assumed that a similar effect of “electron wind” [Fix, 1969] can determine the mobility of dislocations in metals under the action of electric current pulses [Zuev et al., 1978; Boyko, Geguzin, Klinchuk, 1979; Zuev, Sergeev,

1981; Zuev, Gromov, Gurevich, 1990; Lebedev, Hotkevich, 1982; Lebedev, Krylovsky, 1993].

The second aspect of the problem concerns the general nature of the interrelation of auto-wave and dislocation ideas about the development of plastic deformation. As follows from equations (3.9), (3.10) and (4.45), the Taylor–Orowan dislocation equation (1.1.1) is a special case of the more general autowave equation (4.45) and describes only the kinetics of a single deformation act. It can be assumed here that the interaction of dislocations with the formation of dislocation ensembles is a mechanism of self-organization in the defective structure of a deformable medium. In this case, the basis of such a representation may lie both in the model of long-range dislocation and in the model of their close interaction [Indenbom, 1960; Wirtman and Wirtman, 1987; Kuhlmann-Wilsdorf, 2002].

Thus, there is no antagonism between the autowave plasticity model and the dislocation mechanisms involved in explaining it, and the difference in approaches is determined only by the ratio of scales.

The model of generation of autowaves of a localized plastic flow is based on the principle of separation of the active deformable medium into a dynamic and informational subsystem. The quantitative expression of the model is the elastic-plastic deformation invariant connecting the elastic and plastic components of the deformation. The explanation of the nature of the invariant is based on the notion that plastic deformation is a process of self-organization of the defect structure of a deformable medium and is accompanied by a decrease in the entropy of the deformed system.

Analysis of the elastoplastic strain invariant showed that it is performed

- for the stage of linear strain hardening;
- for the stage of easy slip in single crystals;
- for deformation due to the movement of individual dislocations;
- for deformation of the phase transformation.

The physical meaning of the elastoplastic deformation invariant is that plastic and elastic deformations are not just formally connected with each other through the plastic flow curve $\sigma(\varepsilon)$. This relationship determines their mutual development. Information about the nature of such a connection is essential for understanding the nature of the localized plastic flow.

5

A quasiparticle approach in plasticity physics

The use of the autowave model of plastic deformation allows its logical development by the method widely used in condensed matter physics. It consists in the introduction of a quasiparticle corresponding to a given wave process. The universality, fruitfulness and expediency of such an approach are substantiated and demonstrated in the extensive literature on condensed matter physics, for example, in the books of Ziman [1962, 1966], Kittel [1967], Patashinsky, Pokrovsky [1975], Kaganov, Lifshits [1989] and especially Brandt, Kulbachinsky [2007]). The idea of a quasiparticle corresponding to an autowave of a localized plastic flow arises on this background quite naturally.

5.1.On the use of quantum-mechanical ideas in the physics of plasticity

The use of the quasiparticle method in this problem is equivalent to the extension of the principles of quantum mechanics to the physics of plasticity. Such attempts, generally speaking, were made earlier, referring to the three aspects of the problem so far. First of all, it was a quantum-mechanical interpretation of the mechanical properties of crystals. So Feynman [1965] proposed a simple explanation of the nature of the resistance of solids to deformation. In his opinion, a decrease in the volume of the body during compression under the action of external forces causes a decrease in the uncertainty of the electron coordinates x. In accordance with the Heisenberg uncertainty principle $\Delta p \times \Delta x \hbar$ [Landau and Lifshitz, 2004], this leads to an increase in the electron momentum Δp. An interesting attempt to use quantum concepts to explain the instability of the crystal lattice and the nucleation of lattice defects during deformation is also described in Thompson's book [1985].

The second group of works concerned the direct quantum-mechanical interpretation of specific features of deformation, which are not explained in the framework of the traditional approaches of the physics of plasticity. So Bell [1984] drew attention to the possibility of quantizing the elastic moduli of materials, and Steverding [1972] introduced the idea of quantizing elastic waves accompanying the process of destruction. Later this question in an extended interpretation was considered in the work of Maugin [Maugin, 2011 in relation to solitons in an elastic medium. Gilman [Gilman, 1968], and then Oku and Galligan [Oku, Galligan, 1969] attracted the concept of the tunnel effect to explain the separation of dislocations from the attachment points at low temperatures, when the thermal activation mechanism does not work, and evaluated the probability of tunneling. Petukhov and Pokrovsky [1972] gave a rigorous analysis of this phenomenon for the case of dislocation motion in the potential Peierls relief. Recently, Kirichenko et al. [2010] returned to these questions, as well as the authors of [Iqbal, Sarwar, Rasa, 2016].

The third group of studies devoted to the direct introduction of quasiparticles was apparently begun by the fundamentally important work of Morozov, Polak and Fridman [1964]. Its authors, describing the kinetics of the growth of a brittle crack, postulated the existence of a quasiparticle, which they identified with a small zone near the end of the crack moving with it. The quasi-particle was called a *cracon* (from the word crack). This idea was later developed by Morozov, Smirnov [1981], Babkin, Morozov [1981], and Morozov [1998]. In the doctoral dissertation of Miklashevich [2004], the effectiveness of this approach in describing the dynamics of destruction was proved, the mass of the cracon was estimated, and the equations of its motion were derived.

The idea of Morozov, Polak and Fridman [1964] was taken up in later studies of the initial stages of the origin of microdefects. So Zhurkov [1983] introduced the concept of elementary excitation of a crystal – a dilaton, which is a negative density fluctuation. Olemskoi and Petrunin [1987] discussed the possibility of the emergence of a specific strain embryo or fracture, which the authors called frustron (Fig. 5.1 a). An object of this type was observed by Myshlyaev [1972] in a microscopic study of the evolution of the dislocation structure in the course of aluminum creep (Fig. 5.1 b).

Quantization of elementary fracture processes was studied by Slutsker [2005, 2006], and also Gilyarov, Slutsker [2010a, b] in a series of works devoted to clarifying the physical bases of long-term strength in the framework of the approach described by Regel, Slutsker and Tomashevsky [1974]. It was found that the activation volume of the elementary interruption of an interatomic bond in a crystal is close to the atom volume and changes discretely.

As mentioned above, all these attempts lie in the mainstream of one of the most relevant methods of condensed matter physics, based on the introduction and use of quasiparticles. If in the theory of ideal crystals quantum-mechanical models and concepts initiated by the works of Einstein and Debye are widely used (see, for example, Peierls [1956] and Kittel [1967]), then all existing models of plasticity and strength of real crystals with defects are almost entirely based using the principles and methods of classical physics.

5.2. Mass associated with autowave localized deformation

The first attempt to directly use quantum mechanical models to interpret localized plastic flow autowaves was made by Billingsley [Billingsley, 2001]. Applying to the characteristics of autowaves, found in our works [Zuev, Danilov, Kartashova, 1994; Zuev, Danilov, 1997; Zuev, Danilov, 1998], the de Broglie equation written for mass, $m = h\lambda V = 1$, he showed that the value calculated in this way correlates with the atomic mass of the metal under study.

The research we have undertaken to develop the idea of this work included the expansion of the range of materials studied and the correct interpretation of the experimental conditions and autowave characteristics of the deformation process [Zuev, 2004; Zuev, 2005]. If we substitute the experimentally determined values of the lengths λ and the propagation velocity V_{aw} of the phase autowaves for the metals studied to date (Table 5.1) into the de Broglie equation, the empirical values of the effective mass thus calculated

$$m_{ef}^{(emp)} = \frac{h}{\lambda V_{aw}} \tag{5.1}$$

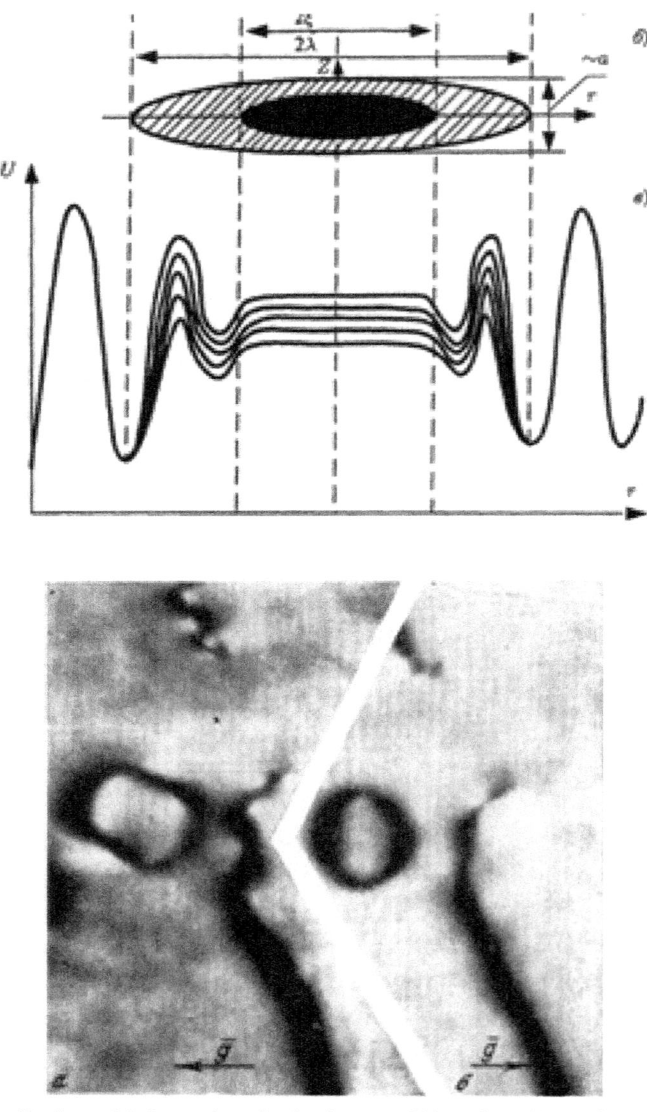

Fig. 5.1. Distribution of deformations in the frustron [Olemskoy, Katsnelson, 2003] (*a*); electron microscopic image of a dislocation micro loop in Al [Myshlyaev, 1972] (*b*).

appear to lie in a rather narrow range $0.3 \leq m_{ef} \leq 4.2$ a.m.u. (1 a.m.u = $1.66 \cdot 10^{-27}$ kg – atomic mass unit). The average effective mass is $m_{ef} = 1.8 \pm 0.3 \approx 2$ a.m.u. The proximity of the values of meth obtained for all metals suggests that the values calculated by the formula (5.1) are not random. The interpretation of the

Table 5.1. Effective mass (a.m.u) calculated from equation (5.1)

	Metals										
	Cu	Zn	Al	Zr	Ti	V	Nb	α-Fe	γ-Fe	Ni	Co
$m_{ef}^{(emp)}$	1.8	1.1	0.5	2.0	1.1	1.4	2.3	1.8	1.8	1.9	1.3

	Metals							
	Sn	Mg	Cd	In	Pb	Ta	Mo	Hf
$m_{ef}^{(emp)}$	1.3	4.0	4.2	1.6	1.3	0.6	0.3	4.0

value of a meph is to consider it as an added mass associated with the accelerated movement of plasticity carriers (dislocations) in a viscous medium [Landau and Lifshitz, 2001]. Such a medium during deformation is the phonon and electron gases of a metallic crystal, whose viscosity was considered in the works of Kaganov, Kravchenko, Natsik [1973] and Alshits, Indenbom [1975a, b]. It is convenient to characterize the mass m_{ef} using the dimensionless parameter $s = m_{ef}/A = A^{-1} \cdot h/\lambda V_{aw}$, introduced in [Zuev, 2005], which is the effective mass calculated by equation (5.1), normalized to the atomic mass A of the metal under study. As it turned out, this parameter is linearly related to the number of electrons per unit cell n by the relation $s = s_0 + \kappa n$ where s_0 and κ are empirical constants. The value of n is a convenient parameter determining the energy spectrum of conduction electrons in a metal crystal [Erokhin, Kalashnikov, 2017]. In accordance with the data of Lifshits, Azbel, Kaganov [1971] and Cracknell, Ward [1978], for metals used in the work $1 \leq N \leq 10$. The graph of this dependence is shown in Fig. 5.2.

The dynamics of dislocation motion during plastic deformation is controlled by the viscous drag force (per unit length) $F \sim BV_{disl}$ determined by the viscosity B of the phonon and electron gases in the crystal [Alshitz, Indenbom, 1975 a, b]. This is true when V_{disl} = const but in the case $V_{disl} \neq$ const and the force F_{vis} should be added the inertial term F_{in}, proportional to the acceleration of dislocation $\dot{V}_{disl} \neq 0$ [Landau, Lifshitz, 2001]. In this case

$$F_{\Sigma} = F_{vis} + F_{in} \approx BV_{disl} + \frac{B}{\nu} \cdot \dot{V}_{disl} = B\left(V_{disl} + \frac{\dot{V}_{disl}}{\nu}\right). \tag{5.2}$$

In metals, the contributions of phonon and electron gases to the coefficient of viscous dragging of dislocations are additive, and it

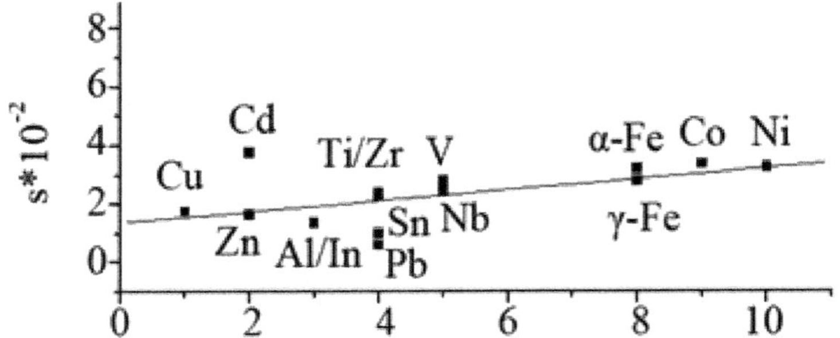

Fig. 5.2. The parameter s as a function of the number of electrons N per unit cell of a metal.

can be assumed that $B = B_{ph} + B_e$ [Suzuki, Yoshinaga, Takeuti, 1989]. In this case, from (5.2) follows

$$F_\Sigma \approx B_{ph} V_{disl} + B_e V_{disl} + \frac{B_{ph}}{\nu}\dot{V}_{disl} + \frac{B_e}{\nu}\dot{V}_{disl}. \qquad (5.3)$$

If v^- is the frequency of deformation acts, then the value B/v, should be interpreted as the added effective mass per unit dislocation length. This coincides with the result of Eshelby [1963], according to which the motion of a dislocation is described by the same equation as the motion of a Newtonian particle. Taking into account the definition of s, we can assume that from equation (5.3) it follows $s \sim B/v \sim m_{ef}$. In this case, the third and fourth terms on the right-hand side of equation (5.3) are associated with the added mass, that is,

$$F_{in} \approx \frac{B_{ph}}{\nu}\dot{V}_{disl} + \frac{B_e}{\nu}\dot{V}_{disl} \approx \frac{B_{ph} + B_e}{\nu}\dot{V}_{disl}. \qquad (5.4)$$

The factor during acceleration V_{disl} is the added mass, which includes the contributions related to the dragging by the phonons (B_{ph}/v) and the electronic (B_e/v) gases. The phonon contribution is only weakly dependent on the nature of the metal, since at the temperature of the higher than the Debye temperature the metal is almost not related to the features of its phonon spectrum. For example, the lattice heat capacity in this temperature range does not depend on the type of substance and temperature (Dulong and Petit law) [Ashcroft, Mermin, 1978]. As for the contribution of the electron gas to the dislocation drag, since, in accordance with the works of Ziman [1962], Kaganov, Kravchenko, Natsika [1973], Suzuki, Yoshinaga, Takeuchi [1989], Lebedev, Hotkevich [1982] and Lebedev, Krylovsky [1993] $B_e \sim N$, so equation (5.4) can be given the form

$$F_{in} \approx \frac{B_{ph} + B_e}{\nu} \cdot \dot{V}_{disl} \approx \left(m_{ph} + m_e\right) \cdot \dot{V}_{disl} \approx \left(s_0 + \kappa \cdot n\right) \cdot \dot{V}_{disl}. \quad (5.5)$$

This interpretation of the effective mass m_{ef} emphasizes the direct connection of the macroscopic characteristics of localized plastic deformation λ and V_{aw} with the parameter of the electronic structure of the metal – the number of electrons in the unit cell. Indirectly, the possibility of the considered mechanism, based on the existence of a mechanical connection of moving dislocations and the electron gas, is confirmed by studies of Bobrov and Lebedkin [1989, 1993], Shibkova et al. [2014]; Shibkov et al., [2016], who connected the electric potential on the surface of a deformable metal sample with the drag of conduction electrons by moving dislocations under sudden plastic deformation.

5.3. Introduction of quasiparticles – autolocalizon

Replace the masses $m_{ef}^{(\mathrm{emp})}$ (Table 5.1) found by equation (5.1) with similar ones in magnitude, but independently calculated from reference mass data $m_{ef}^{(\mathrm{cal})} = \rho r_{\mathrm{ion}}^3$. Here, r_{ion}^3 is the volume per one ion in the crystal lattice (a value on the order of the volume of the Wigner–Seitz cell [Ashcroft, Mermin, 1979]), and r_{ion} is the radius of the ion [Gorelik, Rastorguev, Skakov, 1994]. Then the equation (5.1) can be given the form

$$\lambda \cdot V_{aw} \cdot m_{ef}^{(cal)} = (\lambda V_{aw}) \cdot (\rho r_{ion}^3) \approx \varsigma. \quad (5.6)$$

Equation (5.6) includes two types of quantities. First, these are the characteristics of the phase autowave of the localized plastic flow λ and V_{aw} obtained in the course of the experiments described in Chapter 2. Secondly, it is (independently determined by other researchers) material characteristics of the deformable medium: the density of the substance ρ and the ionic radius r_{ion}.

The results of calculating the value of ζ by the formula (5.6) for all nineteen studied metals are given in Table 5.2. The average value of this value, which is $(6.9 \pm 0.45) \cdot 10^{-34}$ J·s, unexpectedly turns

Table 5.2. Values of ζ calculated by the formula (5.6)

	Metals										
	Cu	Zn	Al	Zr	Ti	V	Nb	α-Fe	γ-Fe	Ni	Co
$\zeta \cdot 10^{34}$	11.9	9.3	2.8	6.1	4.9	3.5	4.9	4.6	4.6	6.1	7.1

	Metals							
	Sn	Mg	Cd	In	Pb	Ta	Mo	Hf
$\zeta \cdot 10^{34}$	8.9	4.9	7.4	9.9	18.4	5.5	3.0	7.3

out to be close to the Planck's quantum constant $h = 6.63 \cdot 10^{-34}$ J·s [Atkins, 1977], and the ratio $\langle\zeta\rangle/h = 1.04 \pm 0.06 \approx 1$.

To understand the meaning of this coincidence, it is necessary to statistically compare the experimentally determined value $\langle\zeta\rangle$ with the reference value of the Planck constant h. The generally accepted statistical procedure used for this purpose consisted in calculating Student's criterion [Mitropolsky, 1961; Hudson, 1967; Stepnov, 2005] according to the formula

$$\hat{t} = \frac{\langle\varsigma\rangle - h}{\sqrt{\hat{\sigma}^2}} \cdot \sqrt{\frac{n_1 \cdot n_2}{n_1 + n_2}}, \tag{5.7}$$

and comparing it with the selected standard criterion value of $\hat{t}_{0.05}$ for a probability of 0.95. In formula (5.7), $n = 19$ is the number of values of ζ obtained from equation (5.4), and $n_2 = 1$. The last condition assumes that Planck's constant is determined in independent experiments with high accuracy, that is, with zero dispersion [Atkins, 1977; Fritzsch, 2009]. Then the combined estimate of the variance of $\langle\zeta\rangle$ and h is calculated as

$$\hat{\sigma}^2 = \frac{\sum_{i=1}^{n_1}\left(\varsigma_i - \langle\varsigma\rangle\right)^2 + \sum_{1}^{1}\left(h - h\right)^2}{n_1 + n_2 - 2} = \frac{\sum_{i=1}^{n_1}\left(\xi_i - \langle\varsigma\rangle\right)^2}{n_1 + n_2 - 2}. \tag{5.8}$$

From formula (5.7), taking into account relation (5.8), it follows that $\hat{t} = 0.046 \ll \hat{t}_{0.05} = 2.13$, where $\hat{t}_{0.05}$ is the value of the Student's criterion for a probability of 0.95 with the number of degrees of freedom $n_1 + n_2 - 2 = 18$.

The result obtained means that with a probability of more than 95% the difference between $\langle\zeta\rangle$ and Planck's constant $\hbar$ is insignificant, that is, the corresponding samples belong to the same general population. Thus, the quantity calculated using relation (5.6) is indeed Planck's constant, and the phenomenon of localized plastic deformation acquires a formal connection with quantum mechanical phenomena.

The next step in the development of the concepts under consideration is the introduction of a quasiparticle, which can be put into correspondence with the autowave process of plastic flow. Following the method adopted in condensed matter physics [Brandt, Kulbachinsky, 2007], we assume that $m_{ef}^{(emp)}$, defined by equation (5.1), is the mass of a hypothetical quasiparticle. The last one in the works of Zuev and Barannikova [Zuev, Barannikova, 2010a, b] was named **autolocalizon**. It represses the phase autowave of a localized plastic flow. The possible characteristics of the autolocalizon introduced on the basis of the idea of its relationship with the phase autowave of localized plastic flow, are listed in Table 5.3.

5.4. Quasiparticle representation of localized deformation

As usual (Brandt, Kulbachinsky, 2007), the introduction of a quasiparticle, an autolocalizon, simplifies the solution of problems associated with explaining the features of the kinetics of plastic flow. Consider the possibility of using a quasiparticle approach to achieve some goals of this kind.

5.4.1. Jump-like plastic deformation

This type of deformation (Portevin–Le Chatelier effect) [Bell, 1984; Kubin, Estrin, 1985, 1992; Kubin, Chihab, Estrin, 1988; Rizzi, Hähner, 2004] is often implemented with plastic flow and is well studied. It manifests itself in the fact that jumps of deformation and sudden drops of deforming stress appear on the plastic flow curve.

Table 5.3. Main characteristics of the quasiparticle–autolocalizon

Characteristic	Formula	Value
Variance law	$\omega(k) \sim 1+k^2$	
Mass (a.m.u.)	$m_{a-l} \equiv h/\lambda V_{aw}$	1.7 ± 0.2
Velocity (m/s)	$V_{a-l} \equiv V_{aw}$	$10^{-5} \dots 10^{-4}$
Pulse (J s/m)	$p = hk = h/\lambda$	$(6 \dots 7) \cdot 10^{-32}$

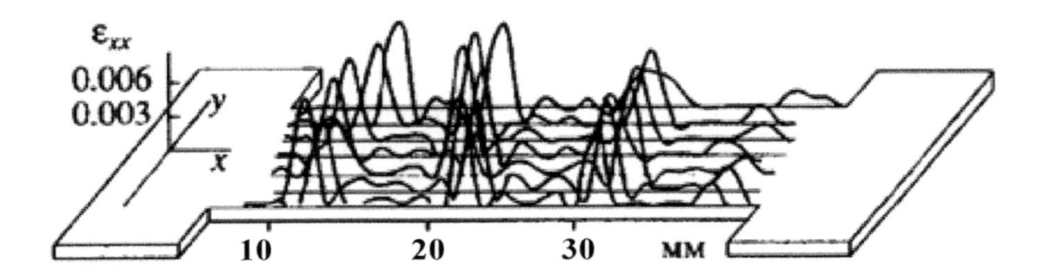

Fig. 5.3. Scheme for determining the number of localization autowaves arising in the sample under tension.

Analyzing this phenomenon, we use the natural assumption that the sample length L, contains an integer $m = 1, 2, 3 ...,$ of the autowaves with length λ, $\lambda = L\ m$ as shown in Fig. 5.3.
We write equation (5.1) in the form

$$\lambda = \frac{h}{\rho r_{ion}^3 V_{aw}} \tag{5.9}$$

and we will assume that since the sample is extended as the strain increases, acquiring a length of $L \approx L_0 + \delta L$, where L_0 is its initial length, then $L/m \approx L_0 m + \delta L/m$. In this case, equation (5.7) implies the relation

$$\delta L \approx \frac{h}{\rho r_i^3 V_{aw}} \cdot m \tag{5.10}$$

and allows one to estimate the value of δL for the elongation jump under the following conditions: $m = 1$, $V_{aw} \approx 3 \cdot 10^{-3}$ m/s, $\rho r_{\rm ion}^3 \approx 1.8$ a.m.u. $\approx 3 \cdot 10^{-27}$ kg. Under these conditions, $\delta L \approx 10^{-4}$ m, which is consistent with the experimentally observed values of the jump length during the Portevin-Le Chatelier effect [Krishtal, 2001, Shibkov et al, 2016; Gorbatenko, Danilov, Zuev, 2017; Zuev, 2017a, b].

Thus, equation (5.10) implies the discreteness of the sample length variation $dL \sim m$, that is, a jump-like deformation, which can be considered as a process of adjusting the sample length to the existing autowave pattern of plastic strain localization. Depending on the nature of the material, the mechanisms and the length of the deformation jumps can be different: eq. (5.10) only establishes that they are obligatory.

From formula (5.10) it also follows that $\delta L \sim V_{av}^{-1}$. The previously obtained data show that the autowave speed is proportional to the speed of movement of the traverse of the testing machine, that is,

$V_{aw} \sim V_m$. In this case, we should expect a decrease in the amplitude of the jump with an increase in the stretching rate. There are convincing experimental confirmations of such a change, presented, for example, by Pustovalov [2008] while studying the spasmodic deformation of aluminium at a temperature of 1.4 K.

5.4.2. Autowave length - autolocalizon displacement length

To estimate these quantities, we use the fact that in the framework of the quasiparticle approach, it is convenient to consider a deformable medium as a mixture of two types of quasiparticles: phonons and autolocalizons. In this case, the quantitative laws of the plastic flow can be explained as a result of the Brownian motion of autololocalizons in a viscous phonon gas, as shown in Fig. 5.4.

In such a case, the characteristic length of the autowave plastic flow localization $\lambda \approx 10^{-2}$ m can be identified with the length of the resulting displacement $R = \left(\sum_{i=1}^{i=n} r_i^2\right)^{1/2}$ with Brownian random motion of the autolocalizon in one period of the autowave process $T = 2\pi/\omega \approx 10^3$ s, equal to n. The magnitude of the displacement R, shown in Fig. 5.4, in this case can be estimated as [Levich, 1969.

$$R \approx \sqrt{\frac{k_B T}{\pi B r_{a-l}}} \cdot \tau \approx \sqrt{\frac{2k_B T}{\omega B r_{ion}}} \tag{5.11}$$

If we assume that the size of the autolocalizon is $r_{a-1} \approx r_{\text{ion}} \approx 10$–$10$ m, and the dynamic viscosity of the phonon gas is $B \approx 10^{-4}$ Pa·s,

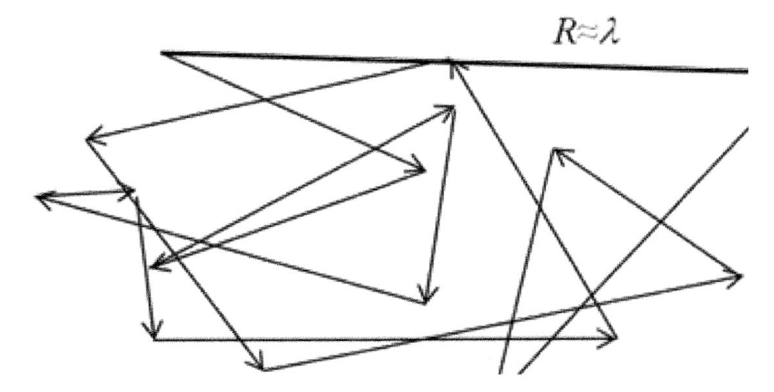

Fig. 5.4. Autowave length (bold line) as a result of random walks of a Brownian particle (autolocalizon) over a period of localized flow autowave.

then at $T = 300$ K the displacement is $R \approx 10^{-2}$ m, that is, coincides with theautowave length of localized deformation λ. In addition, from equation (5.11), written in the form

$$\frac{R^2\omega}{2\pi} \approx \frac{k_B T}{\pi B r_{a-l}}, \tag{5.12}$$

where $R^2\omega/2\pi = 1.3 \cdot 10^{-7}$ m²/s. Then, the equality $R^2\omega/2\pi \approx D_{\varepsilon\sigma}$ is possible, indicating the connection between the elastoplastic deformation invariant and the quasiparticle notion of plastic flow.

5.4.3. Elastoplastic deformation invariant and autolocalizon

If we simultaneously use the equation of the elastoplastic invariant (4.11) and equations (5.1) and (5.4) for the effective mass, then it is possible to calculate the ratio, the value of which is equivalent to the elastoplastic invariant deformation

$$\hat{Z}_{calc} \equiv \frac{\lambda V_{aw}}{\chi V_t} = \frac{h}{\rho r_{ion}^3 \chi V_t}. \tag{5.13}$$

The quantity $\hat{Z}_{calc}$ calculated in such a way is determined only by the material (lattice) constants of the medium. The results of the calculation according to equation (5.13) are given in Table 5.4. The average value of the calculated values is $\langle \hat{Z}_{calc} \rangle = 0.63 \pm 0.1 \approx 1/2$, and the ratio $\langle \hat{Z} \rangle_{exp} / \langle \hat{Z} \rangle_{calc} = 0.97 \approx 1$, that is, almost coincides with the experimentally observed value. This coincidence serves as proof of the validity of the invariant (4.11).

5.5. Spectrum of elementary excitations of a deformable medium

As was shown above, deformation processes developing in a deformable medium include interconnected elastic (wave) and plastic (autowave) components, on the interaction of which a two-component model of localized plastic flow considered in Chapter 4 is constructed. Both components are characterized by corresponding dispersion ratios. The close relationship between the elastic and plastic components of the deformation (see Chapter 4) allows us to hope that we can obtain a general dispersion law for elastic and plastic deformations.

Table 5.4. Comparison of calculated and experimentally determined values of $\hat{Z}$

	Metals										
	Cu	Zn	Al	Zr	Ti	V	Nb	α-Fe	γ-Fe	Ni	Co
$\hat{Z}_{calc}$	0.75	0.31	1.05	0.55	0.32	0.45	0.33	0.54	0.34	0.35	0.5
$\hat{Z}_{exp}$	0.42	0.5	4.47	0.33	0.42	0.85	0.44	0.46	0.48	0.35	0.38

	Sn	Mg	Cd	In	Pb	Ta	Mo	Hf
$\hat{Z}_{calc}$	0.65	0.63	0.27	1.18	1.4	2.7	0.4	0.33
$\hat{Z}_{exp}$	0.48	0.98	0.24	0.78	0.5	1.1	0.2	0.24

5.5.1. Hybridization of the spectra of an elastically and plastically deformable medium

A two-component model linking acoustic and deformation processes in solids allows for the possibility of hybridization of the spectrum of elementary excitations of an elastic (phonon) and deformation (autolocalizon) subsystems of a deformable crystal. Such a spectrum can be obtained by combining a linear graph of the dispersion relation for transverse phonons $\omega \approx V_t \cdot k$ (without taking into account the dispersion of the speed of sound in the high-frequency region at $\omega \rightarrow \omega_D$) [Raceland, 1975; Ashcroft, Mermin, 1979; Lüthi, 2007] with a parabolic dispersion relation for autowaves of localized plasticity (autolocalizons) $\omega = \omega_D + \alpha(k-k_0)^2$ (eq. (3.27)). The result of such a combination for polycrystalline aluminium is shown in Fig. 5.5 a.

The estimates show that the coordinates of the intersection point of the graphs in the high-frequency region (inset in Fig. 5.4 a) are physically justified: $\hat{\omega} \approx \omega_D \approx 10^{13}$ Hz (ω_D is the Debye frequency), and the wavenumber $\hat{k} = 2\pi/\hat{\lambda}$ corresponds to the minimum possible wavelength $\lambda_{min} = \hat{\lambda} \approx \chi$. This indirectly confirms the applicability of the description of plastic flow as the interaction of a phonon gas with quasiparticles corresponding to autowaves of a localized plastic flow. More interesting is the situation near the intersection point of the linear and parabolic dependences in the low-frequency region for small values of the wave number, where, obviously, the hybridized dependence $w(k)$ or $E(p)$ has a local maximum. Its existence can be associated with the creation of dislocations during elastic nonlinear deformation of an initially defect-free crystal to the mechanics of the condensation of long-wave phonons proposed by Umezzaw,

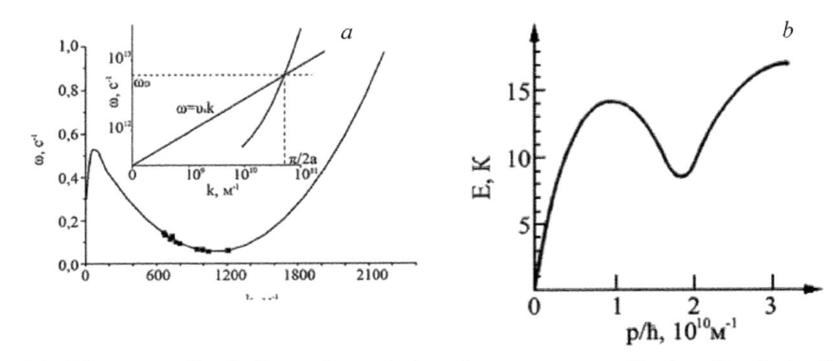

Fig. 5.5. The generalized dispersion relation for autowaves of a localized plastic flow $\omega \sim 1 + k^2$ and elastic waves $\omega \sim k$; the inset shows the high-frequency region of the dispersion relations (*a*); dispersion curve for superfluid He^4: energy in Kelvin $E = \hbar\omega/k_B$, wavenumber in units of $k = p/\hbar$ [Brandt, Kulbachinsky, 2007] (*b*).

Matsumi and Tatik [1985]. The individual dislocations that appear as a result of this process are not resistant, but their energy decreases with increasing or decreasing wave number k. It can be assumed that the first case corresponds to the formation and development of low-energy dislocation structures [Kuhlmann-Wilsdorf, 2002 (in the limit, regions or autowaves of localized plastic deformation), and the second leads to the emergence of fragile microcracks [Zuev, 1998].

According to the dispersion relation for autowaves of localized plastic flow (3.22), there is a $\theta \leq w_0 \approx 10^{-2}$ gap in the vibration spectrum $\omega(k)$. As at any temperature $\hbar\omega_0 \ll k_0T$, spontaneous localization of plastic deformation should be excited at arbitrarily low temperatures. Indeed, the jump-like deformation associated with the localization of the flow was repeatedly observed at $T \leq 1$ K [Pustovalov, 2008], and the lack of localization of plastic deformation is possible only because of geometric constraints for small sample sizes, which was discussed above in connection with the scale effect for the autowave localized plastic flow.

When analyzing the hybridized dependence $\omega(k)$ for plastic flow, shown in Fig. 5.5 a, it is easy to catch its similarity with the dispersion curve for superfluid helium [Kaganov, Lifshits, 1989; Landau, Lifshits, 2002; Brandt, Kulbachinsky, 2007], shown in Fig. 5.5 b. The minimum on the parabolic branch of this curve corresponds to the production of rotons – elementary excitations of a medium with a quadratic, and not linear, like that of phonons, law of dispersion. The formation of the rotons is responsible for the superfluidity of liquid helium. On the basis of the similarity of the forms of dispersion curves, it can be assumed that the autolocalizon as a quasi-particle is an analogue of the roton born during localized plastic deformation and determines the kinetics of this process.

5.5.2. Dispersion and effective mass of autolocalizon

Using the dispersion relation (3.22) for autowaves of localized plasticity, it is possible to calculate in a standard way [Brandt, Kul'bachinsky, 2007] the effective mass of autolocalizon corresponding to the autowave of a localized plastic flow,

$$m_{a-l} = \hbar \frac{\partial^2}{\partial k^2}\left[\omega(k)\right], \tag{5.14}$$

which is 0.1 a.m.u. for Al and 0.6 a.m.u. for Fe. Effective masses calculated by the formula (5.1) $m_{ef} = h/\lambda V_{aw}$ and are given in Table 5.1 for aluminium and iron are 0.5 and 1.8 a.m.u, respectively, that is, they have the same order of magnitude. These estimates confirm the validity of the assumption that the quasiparticles (autolocalizons) with an average mass of ~ 1.7 a.m.u. correspond to the wave processes of localization of plastic deformation. According to Brandt and Kulbachinsky [2007], the effective mass of the roton is 0.64 a.m.u. The similar proximity of the effective masses of two quasiparticles also testifies in favour of the hypothesis under discussion.

5.5.3. Condensation of quasiparticles in the process of plastic flow

There is another interesting opportunity to clarify ideas about the nature of the elastoplastic deformation invariant. Equation (4.11), written relative to the mass, has the form

$$\frac{h}{\lambda V_{aw}} = 2\frac{h}{\chi V_t}, \tag{5.15}$$

where $\frac{h}{\lambda V_{aw}} = m_{a-l}$ is the de Broglie autolocalison mass, and $\frac{h}{\chi V_t} = m_{ph}$ is the phonon mass. In this case, equation (5.13) reduces to the equality $m_{a-1} \approx 2m_{ph}$, which can be considered as a condition for the birth of a new quasiparticle (autolocalizon) with a small momentum

$$p_{a-l} = \left(m_{a-l}V_{aw}\right) = \left[2\left(m_{ph}V_t\right)^2\left(1+\cos\alpha\right)\right]^{1/2} = \sqrt{2}m_{ph}V_t\left(1+\cos\alpha\right)^{1/2}, (5.16)$$

as shown in Fig. 5.6. Here, 2α is the angle between the phonon pulses.

The probability of such a process was first calculated by Landau and Khalatnikov [1949] in explaining the production of rotons in superfluid helium. Details of the combination of two phonons were described by Raceland [1975]. The process under discussion is close to that considered by Umedzawa, Matsumoto and Tatika [1985], who introduced the quantum mechanism of defect formation and showed that the formation of dislocations and their ensembles (walls, grain boundaries and other defects) in an ideal crystal is the result of Bose condensation of long-wave phonons during deformation .

The above reasoning creates the basis for the development of new ideas about the problem of the staging of a localized plastic flow of solids. These representations consist in the fact that quasi-particles of a certain type correspond to each stage of deformation. Such a point of view was first formulated by Vladimirov [1987] in the following form: "*This situation is common to any quasiparticles, the strong interaction between which leads to their collective effects and streamlining. . Defects of the renormalized vacuum are quasi-particles of the following order.*"

A typical example of similar condensation of quasiparticles during plastic flow is the collapse of the autowave of a localized plastic flow described in Section 2.4, observed at the pre-fracture stage of metals and alloys (see Fig. 2.18). This process can be seen as the birth of a cracon, that is, the beginning of the process of destruction.

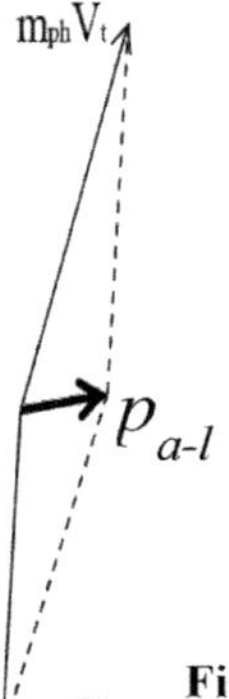

Fig. 5.6. Scheme of autolocalizon birth in a two-phonon process. Thick arrow - the momentum of a formed autolocalizon.

Experimental data allow us to qualitatively present the process of generating structural defects as follows. At the stage of elastic deformation in the crystal, there are only elementary excitations – phonons, which redistribute the elastic stresses in the crystal lattice. In condensation of phonons, in accordance with the mechanism proposed by Umdzawa, Matsumi and Tatika [1985], the crystal produces plastic shear (dislocation) quanta, thereby giving rise to plastic flow [Katanaev, 2005]. Intergenerational stages can occur when dislocations are generated at which defects such as the above-mentioned dilatons [Zhurkov, 1983], frustrons [Olemskoy, Petrunin, 1987] or so-called precursors of dislocations [Psakhie, Zolnikov, Kryzhevich, 2007; Psakhie et al., 2015] form. The evolution of dislocation ensembles, including chaotic distributions of dislocations, their flat clusters, cellular, fragmented and band substructures, has already been considered as a process of condensation of defects [Kozlov, Starinchenko, Koneva, 1993]. At the macroscale level, this process is manifested in the formation of autowaves of localized plastic flow and includes a series of successive stages of the restructuring of autowave structures. Finally, the collapse of the autowave of localized deformation can be considered as the condensation of localized deformation regions with the formation of a quantum of destruction – cracon and its further movement during destruction.

5.5.4. The general meaning of the introduction of autolocalizon

The striking closeness of the quantity $\langle\zeta\rangle$ calculated by equation (5.6), containing the macroscopic experimentally determined quantities λ and V_{aw}, to the reference value h, discussed above indicates a direct connection of the macroscopic phenomena of plastic flow localization with lattice properties. This conclusion may have an explanation, based on the concept of universality [Imrie, 2002], according to which various measurable values do not depend on the majority of the microscopic properties of the system (see also Dotsenko's work [2011]). Prigogine [2005] spoke in the same spirit and wrote: '*When dealing with large systems, we can ignore the fluctuations arising in them and limit them to a description on a macroscopic level*''. Applied to the process of localized plastic flow, this means that the space-time dislocation microstructural heterogeneity of a deformable medium to a lesser extent determines the measurable macro-mechanical characteristics, such as flow stress, yield strength, strain hardening rate and type of pattern of plastic flow localization than is assumed in the framework of dislocation models.

This is also indicated by the data of the already discussed work of Kozlov, Starenchenko, and Koneva [1993], in which it is shown that as deformation increases, dislocation substructures of certain types stop in their development, and then completely disappear, being replaced by new dislocation ensembles. In this case, the slope of the flow curve $\sigma(\varepsilon)$ almost does not change.

Thus, the indicators of mechanical properties are largely controlled by the lattice characteristics of the elastic medium and, accordingly, the laws of quantum mechanics. From this point of view, the obvious importance of the relation (5.6), connecting the quantities traditionally considered heterogeneous, is additionally confirmed. The existence of such a connection is usually neglected in the development of various models of plastic flow, but the relation (5.6) clearly indicates the illegality of this neglect.

5.5.5. Plasticity as a macroscopic quantum phenomenon

The consistent development of such ideas suggests the possibility of attributing localized plastic flow to a number of macroscopic quantum phenomena, such as superfluidity, superconductivity [Pitaevsky, 1966], and the quantized Hall effect [Klitzing, 1986; Dolgopolov, 2014]. In these phenomena, the quantum properties of matter appear on a macroscopic scale (Table 5.5), and the macroscopic equations that describe these phenomena contain Planck's constant.

Formally, the possible quantum nature of plastic flow is indicated by the analogy between the forms of dispersion curves for plastic autowaves (Fig. 5.5) and for superfluid helium. In addition, the quadratic-dispersion curve of autowaves of localized plasticity is similar to the spectrum of elementary excitations of a superconductor [Brandt, Kul'bachinskii, 2007].

The similarity of the physical nature of these macroscopic phenomena is that the underlying effects do not allow the description within the framework of models based on the additive properties of individual carriers. In the case of plastic flow in this quality, dislocations can be discussed, each of which can be considered a shift quantum [Katanaev, 2005]. For this reason, to correctly describe plastic flow, it is necessary to take into account the phenomena of self-organization of a deformable medium, which are realized in the

Table 5.5. Comparison of macroscopic quantum phenomena

Phenomenon	Quantized characteristic	
	Value	Formula
Superconductivity (Brandt, Kul'bachinskii, 2007)	Magnetix flux	$\Phi = \frac{\pi \hbar c}{e} \cdot m$
Superfluidity (Brandt, Kul'bachinskii, 2007)	Speed of rotation of vortices in superfluid helium	$v = \frac{\hbar}{A_{\mathrm{He}}} \cdot \frac{1}{r} \cdot m$
Quantized Hall effect (Klitzing, 1986)	Hall resistance	$R_H = \frac{h}{e^2} \cdot \frac{1}{m}$
Strain jumps in Portevin-LeChatelier effect	Elongation at a strain jump	$\delta L = \frac{h}{\rho r_{ion}^3 V_{aw}} \cdot m$
Comment: e – electron charge; c – speed of light; r – radius of the vortex in superfluid helium, m = 1, 2, 3 ...		

form of autowaves of localized plastic deformation, associated, as shown above, with various strain hardening mechanisms described by Seeger [1960] and Mughrabi [Mughrabi, 1983, 2001, 2004], Vladimirov and Kusov [1975], Vladimirov [1987], Argon [2008] and others.

We now give some additional considerations in favour of the considered formal similarity of plastic flow with superfluidity. We note first that the Indenbom [1979] drew attention to the analogy between dislocations and quantized vortices in superfluid helium or quantized currents in type-II superconductors. Besides, in a deformable material, as in superfluid helium, two types of motions coexist. On the one hand, these are slow movements of individual volumes during the shape change of the body as a whole, and on the other hand, the movement of dislocations with high speeds, which ensures this change. The first of them corresponds to a high viscosity $\eta \approx 10^6$ Pa · s, and to the second, the viscosity of the phonon gas $B \approx 10^{-4}$ Pa · s [Alshitz, Indenbom, 1975 a, b]. Thus, the elements of the deformable medium move at different speeds. Since, in viscous motion, $VF_1B = 1$ or VF_1/η, where $F_1 = \sigma L$ is the Pitch–Köhler force [Friedel, 1967] and $B \ll \eta$, it is possible to estimate these velocities. So, for example, for an autowave of switching on a yield plateau of

$\sigma \approx 100$ MPa, and $L \approx \lambda \approx 10^{-2}$ m. Estimation of the front movement speed is consistent with what is observed in experiments [Krishtal, 2001; Shibkov, Zolotov, 2009; Shibkov, Zolotov, Yellow, 2012; Shibkovetal, 2016; Gorbatenko, Danilov, Zuev, 2017] the value is $V_{aw} \approx 10^{-5}$ m/s. On the other hand, using low viscosity, we obtain $V_{disl} \approx 10^{2}$ m/s, which also corresponds to the experimentally observed speed of movement of dislocations during their above-barrier movement in a crystal [Nadgorny, 1972; Lubenets, 1973; Alshits, Indenbom, 1975a, b; Suzuki, Yoshinaga, Takeuchi, 1989].

Such an approach, based on quantum-mechanical concepts, is perhaps surprising for specialists in the field of plasticity physics, since the spatial scale of macroscopic phenomena of plasticity considerably exceeds the scales for which the quantum approach is traditionally used. So the ratio of wave ($\lambda \approx 10^{-2}$ m) and dislocation ($b \approx 10^{-10}$ m) scales $\lambda/b \approx 10^{8}$. However, at the dislocation level, quantization of deformation seems natural, since, due to the discrete crystal lattice, the Burgers vector b can be interpreted as a quantum of shear deformation, as indicated, for example, by Umedzawa, Matsumoto, Tatiki [1985] and Katanaev [2005]. Thus, the use of the "wave–quasiparticle" dualism to describe the processes of localization of deformation is fully justified.

5.6. Deformation localization and periodic table of elements

It can be expected that additional information on the relationship between the plasticity of metals and their position in the periodic table of elements would be useful for progress in understanding the nature of plastic deformation. Such a relationship has been studied in detail earlier when analyzing the correlation of a number of physical properties of elements, such as the Debye temperature, binding energy, density, melting point, elastic modulus, electron work function, and others, with the atomic number of elements Z [Grigorovich, 1966; Grimwald at al., 2012].

5.6.1. General characteristics of the problem

As was shown in Section 4.3, the characteristic scales λ and χ and the corresponding propagation velocities V_{aw} and V_t form a dimensionless ratio $\lambda V_{aw}/\chi V_t = \hat{Z} =$ (4.11), called the elastoplastic deformation invariant. As a characteristic of the development of the autowave itself of a localized plastic flow, it is advisable to use the complex plasticity parameter $\lambda \approx V_{aw}$ with the kinematic viscosity dimension $L^2 \cdot T^{-1}$ [Zuev, 2011]. Obviously, this value combines the spatial characteristic of the localization pattern λ and the propagation speed of the autowave of localized plasticity V_{aw}.

Of great interest may be the connection of these characteristics with the electronic structure of the elements studied. In a fairly simple approximation, the corresponding data are determined from the periodic table of elements. We will keep in mind that the period number N in it coincides with the number of electron shells of atoms, and the number of conduction electrons in the unit cell n for all the metals studied, except for Fe, Co and Ni, is determined by the group number. A detailed analysis of the characteristics of the localization patterns was carried out for nineteen different metals located in the 3rd, 4th, 5th and 6th periods of the periodic table of elements, as summarized in Table 5.6. Additionally, the bottom line shows the number of conduction electrons n per unit cell of the metal, taken from the monograph by Cracknell and Wang [1978].

5.6.2. Experimental data

Analysis of the obtained data allowed to establish that, at least within 12 (Mg) $\leq Z \leq$ 82 (Pb) range, the plasticity parameter λV_{aw} oscillates about the mean value. These oscillations correspond to similar laws of the behaviour of a number of independently determined lattice characteristics. This correspondence is illustrated as shown in Fig. 5.7 by the periodic behaviour of the dependences of λV_{aw} and the Wigner–Seitz cell size [Newnham, 2005] on the atomic number of the element Z. At the same time, the experiment shows that the plasticity parameter λV_{aw} correlates with a number of other physical properties, as shown in Fig. 5.8 a–d. In these cases, the corresponding dependences are split in accordance with the numbers of the periods to which the studied metals belong.

When analyzing the behaviour of the parameter λV_{aw} within the limits of each period of the Periodic Table of the Elements, the data of Fig. 5.2 were used. Their processing made it possible to establish that $(\lambda V_{aw})^{-1} \sim n$, for each of the studied periods of the Periodic Table, the ratio

$$(\lambda V_{aw})^{-1} \approx C + Dn, \tag{5.17}$$

Table 5.6. Investigated metals and elastoplastic deformation invariant

Periods	Rows	Metals Groups of the periodic table of elements									$\frac{\lambda \cdot V_{aw}}{\chi \cdot V_t} = \hat{Z}$	
		I	II	III	IV	V	VI	VIII				
3	III		$_{12}$Mg	$_{13}$Al							$\langle \hat{Z} \rangle_3 = 0.57 \pm 0.63$	
4	IV				$_{22}$Ti	$_{23}$V		$_{26}$Fe	$_{27}$Co	$_{28}$Ni	$\langle \hat{Z} \rangle_4 = 0.50 \pm 0.15$	
	V	$_{29}$Cu	$_{30}$Zn									
5	VI				$_{40}$Zr	$_{41}$Nb	$_{42}$Mo				$\langle \hat{Z} \rangle_5 = 0.48 \pm 0.15$	
	VII		$_{48}$Cd	$_{49}$In	$_{50}$Sn							
6	VIII				$_{72}$Hf	$_{23}$Ta					$\langle \hat{Z} \rangle_6 = 0.69 \pm 0.45$	
	IX				$_{82}$Pb							
8		1	2	3	4	5	6	8			9	10

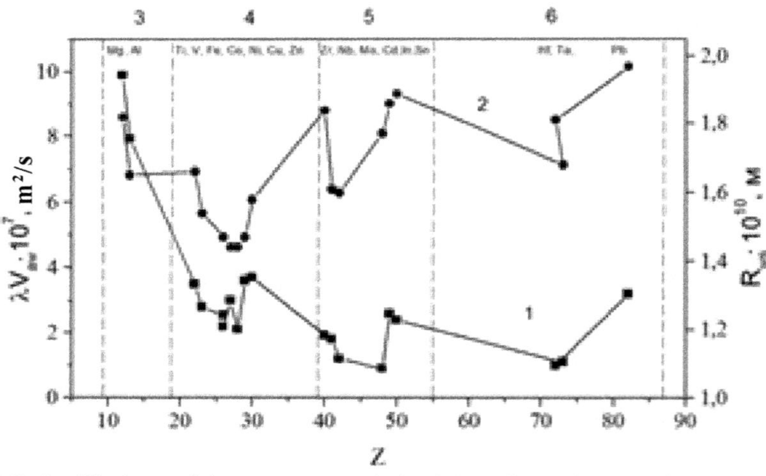

Fig. 5.7. Oscillations of the parameter of plasticity of metals (1) and the radius of the Wigner-Seitz cell R_{WS} (2) depending on the atomic number of the element Z.

in which the coefficients C and D are different for elements of the periods 3, 4, 5 and 6, Processing the experimental data showed that the coefficient D in equation (5.17) depends exponentially on the value of N (Fig. 5.9)

$$D \approx D_0 \exp(-q/N), \tag{5.18}$$

where N = 4, 5, 6 is the number of the period in the periodic table of the elements, and D_0 and q are empirical constants. The correlation coefficient of the quantities are $\ln D$ and $N^{-1} \sim (-1)$, which corresponds to the functional connection in the form of the relation (5.18).

Thus, during the deformation of a solid, the autowave characteristics of the process are associated with the electronic structure of the metals, that is, with the position occupied by the metal in the periodic table of elements. This relationship manifests itself as a complex dependence of the macroscopic characteristic of the development of autowave localized plasticity λV_{aw} on the numbers of groups n and periods N in the Periodic Table of lements.

5.6.3. Interpretation of the data

Let us compare the averaged values of the invariant $\langle \hat{Z} \rangle_i$ and for different periods, given in Table 5.6, and also discuss the nature of the dependence of the plasticity parameter λV_{aw} on the electron density of the investigated metals, that is, on the position of the elements in the Periodic Table of Elements.

As follows from Table 5.6, the experimentally estimated values of the invariant $\langle \hat{Z} \rangle$ and for the 3rd–6th periods of the periodic table are somewhat different from each other. To check the significance of the difference, a statistical procedure was used for comparing the mean values of the invariant $\langle \hat{Z} \rangle$ and for these periods using the Student's t-test using the relation (5.7).

Comparison of t-criteria for the compared pairs showed that the difference in the values of $\langle \hat{Z} \rangle_3$, $\langle \hat{Z} \rangle_4$, $\langle \hat{Z} \rangle_5$ and $\langle \hat{Z} \rangle_6$ is statistically insignificant, so with a probability of ~0.85 these values belong to the same general population. Averaging over all metals gives $\langle \hat{Z} \rangle_{Me} = 0.55 \pm 0.10 \approx 1/2$.

Thus, we can assume that the value of the elastoplastic strain invariant does not depend on the number of the period in which the metal under study is located in the periodic table of elements. The invariant is a universal characteristic of the process of developing a localized plastic flow at the stage of linear strain hardening.

As calculations show, for metals of the 3rd-6th periods $(\lambda V_{aw})^{-1} \sim n$. When analyzing this pattern, one should pay attention to the relationship between the quantity $(\lambda V_{aw})^{-1}$ and the effective mass calculated using formula (5.1). In this case, $m_{eff} \sim n$. In addition, experimental data have shown that the dependence which holds for metals of the 3rd to 6th periods is the one shown in Fig. 5.10 $\ln(\lambda V_{aw})^{-1} - \chi^{-1}$, and the correlation coefficients of the values $(\lambda V_{aw})^{-1}$ and $\chi^{-1}$are $R_3 = -1$, $R_4 = -0.52$, $R_5 = -0.61$ and $R_6 = -0.6$, respectively.

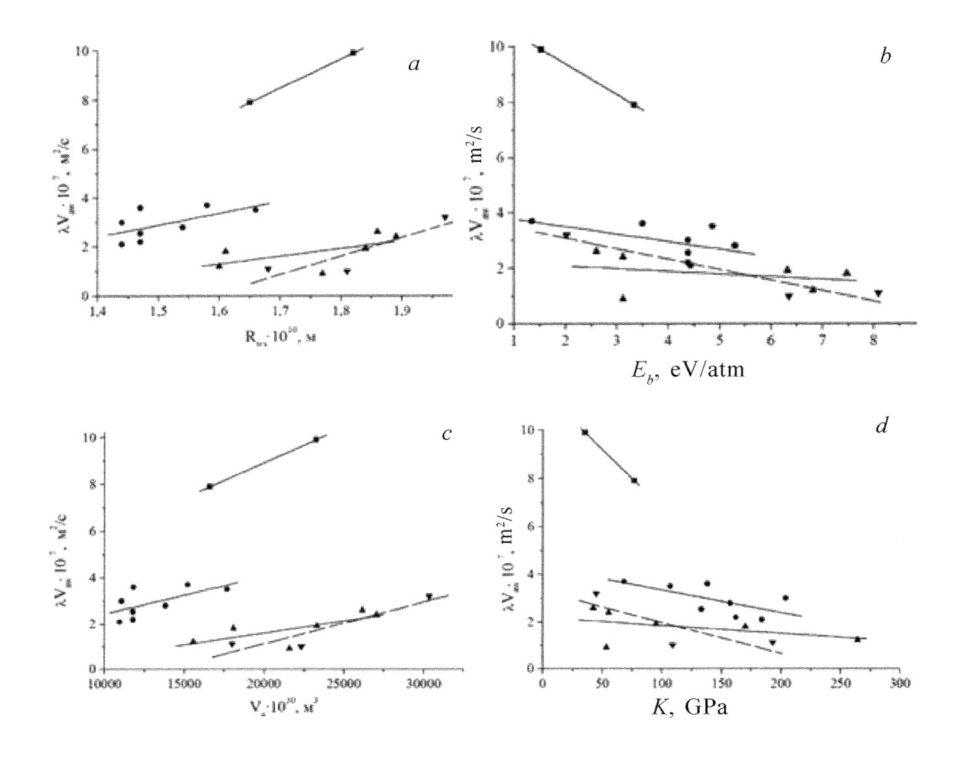

Fig. 5.8. Connection of the plasticity parameter with the Wigner-Seitz cell radius (*a*), binding energy (*b*), volume per atom (*c*), bulk elastic modulus (*d*). ■ – elements of the 3rd - period (Mg, Al), ● – elements of the 4th period (Ti, V, α-Fe, γ-Fe, Co, Ni, Cu, Zn), ▲ - elements of the 5th period (Zr, Nb, Mo, Cd, In, Sn); ▼ – elements of the 6th period (Hf, Ta, Pb). Correlation coefficients: (*a*) R = 1; 0.6; 0.6; 0.9; (*b*) R = 1; 0.53; 0.4; 0.95; (*c*) R = 1; 0.61; 0.72; 0.92; (*d*) R = –1; –0.7; –0.5; –0,8.

Then inside the period the following ratio holds

$$(\lambda \cdot V_{aw}) \sim \exp\left(\frac{\chi_N^*}{\chi}\right), \tag{5.19}$$

where the constant χ_N^* for periods with numbers N = 3, 4, 5, 6 can be found from experimental data. As it turned out, this value is close to the interplanar spacing χ_{in} for alkali metals Li, K, Rb and Cs, from which the 3rd, 4th, 5th and 6th periods begin, respectively. Numerical estimates show that for each of the periods $\chi_N^* \approx 1.5\chi_{in}$. The discussed regularity corresponding to the relation (5.19) is important because it can be used to predict the plastic flow characteristics of metals

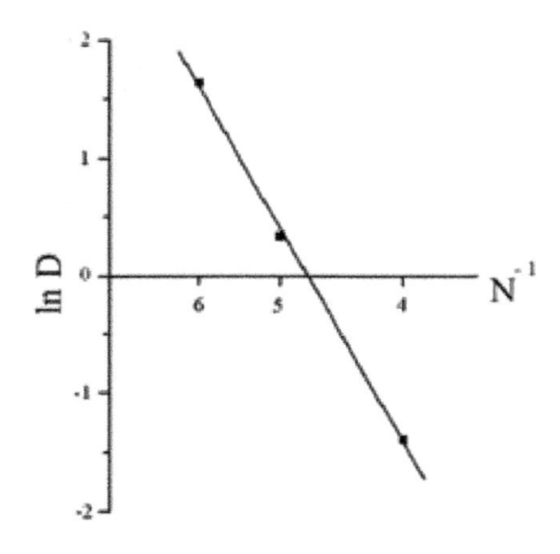

Fig. 5.9. The dependence of the coefficient D in equation (5.18) on the period number N. The correlation coefficient $R = -1$.

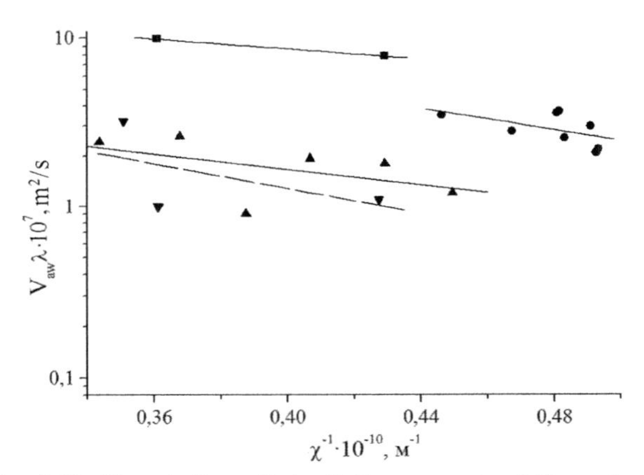

Fig. 5.10. Correlation of plasticity parameter ΔV_{aw} and value χ^{-1}.

by their lattice parameters.

A likely cause of the manifestation of quantum effects in macroscopic plastic deformation processes can be the close connection of plasticity effects with the characteristics of the crystal lattice, as reflected in the two-component model proposed in Chapter 4. In other words, the generation of autowaves of a localized plastic flow – the self-organization of a deformable medium – is realized in an elastic medium, the behaviour of which is governed by quantum mechanical laws. Therefore, the processes of energy redistribution during the generation of autowaves of localized plasticity in a system capable of self-organization turn out to be subordinate to the quantum nature of crystals.

Experimental results and their interpretation also indicate the need to take into account the close relationship of defect ensembles with the phonon subsystem in crystals. The correct description of such a relationship is possible both in terms of the interaction of wave and autowave processes, and in terms of the interaction of quasiparticles – phonons and autolocalizons. Both of these variants satisfactorily describe the two-component model of plastic flow developed on the basis of such representations and correctly explain the occurrence of the macroscopic scale of the observed regularities of plastic flow localization in deformable metals and alloys.

Conclusion

The developed approach to the problem of the development of plastic deformation of solids puts macroscopic – autowave — laws of development of localized plastic deformation at the forefront. Within its framework, the localization of plastic the flow is considered as an integral and, perhaps, the most important and informative attribute of this process, resulting from the interaction of the elements of the defect structure of a deformable medium. The formation of large-scale space-time localization patterns – patterns of localized plastic flow correlated with the stages of strain hardening of materials – is interpreted as acquiring a certain degree of order by a deformable system, and the deformable body itself is viewed as an open thermodynamic system whose evolution is controlled by its own state and energy flow from loading device. Within the framework of this approach, the theory of plasticity acquires the right and possibility to use the concepts and apparatus of synergetics.

Synergetic ideas about the emergence of order in open systems led to the creation of a two-component model of localized plastic flow. It is based on a causal relationship between the elements of two subsystems of a deformable medium, one of which combines acts of plasticity of a different nature, and the second – acoustic emission signals arising in this case. The mathematical form of the model, the elastoplastic invariant of deformation, makes it possible to obtain a consistent and fairly intuitive method for the macroscopic description of the plastic flow. This method is suitable for cases of deformation, realized by different mechanisms, is performed for materials of different physicochemical nature and is therefore sufficiently universal. Many important regularities of plastic flow found explanation as a consequence of the elastoplastic invariant.

What, in the author's opinion, was it possible to do in this work?

First of all, it was possible to build a unified system of views on the nature of a macroscopically localized plastic flow, taking into account such properties of deformable media as their nonequilibrium, nonlinearity, activity, and memory.

It was also possible to quantitatively describe the multilevel nature of the processes in a deformable medium. The proposed variant of the description covers the scale of the crystal lattice, dislocation, mesoscopic and macroscopic processes, setting the quantitative ratio between the spatial characteristics of the deformation processes.

And finally, it was possible to find a natural connection between the autowave theory of plastic deformation and the theory of dislocations and to show that the latter is the limiting case of the first and is suitable for deformation of materials with a low dislocation density.

Of course, I know straight away that many aspects of the plastic flow have not been explained. These aspects can be listed and should serve as research topics in the coming years.

First of all, it concerns the elucidation of the mechanisms and regularities of a multiple change of autowave deformation modes during the stretching of samples with a constant speed.

It was not possible to solve the autowave equations introduced in Chapter 3 of the processes of plastic flow, having previously written them in a form that would take into account the plausible form of the point stress kinetics and strains in them.

Finally, it has not yet been possible to bring to the technological applications the patterns of plastic deformation obtained in the work and the related fundamental possibilities of predicting the behaviour of materials under load and evaluating technological plasticity.

Each of these unresolved problems deserves close attention and may serve as a topic for further research of various, unfortunately, to a greater degree of high complexity.

Bibliography

Adamson A. Physical surface chemistry. - Mosow, Mir, 1979. - 568 p.

Alshits V.I., Indenbom V.L. Dynamics of dislocations. Problems of modern crystallography. - Moscow, Nauka, 1975a. - S. 218–238.

Alshits V.I., Indenbom V.L. Dynamic braking of dislocations. Usp. Fiz. Nauk. - 1975b. - T. 115. - No. 1. - S. 3–39.

Alshits V.I., Darinskaya E.V., Koldaeva M.V., Kotovsky R.K., Petrzhik E.A., Tronchik P. Physical kinetics of dislocation motion in non-magnetic crystals: a view through a magnetic window. Uspekhi Fiz. - 2017. - T. 187. - No. 3. - S. 327–341.

Amelinks S. Methods of direct observation of dislocations. - Moscow, Mir, 1968 .- 440 p.

Ango A. Mathematics for electrical and radio engineers. - Moscow, Nauka, 1967 .- 779 p.

Anderson O. Definition and some applications of isotropic elastic constants of polycrystalline systems obtained from data for single crystals. Physical Acoustics. T. IIIB. - Moscow, Mir, 1968. - S. 62–121.

Andronov V.M., Gvozdikov A.M. Stress state at the front of the Luders-Chernov band and unstable plastic flow. Fiz. Met. Metalloved. - 1987. - V. 63. - No. 6. - S. 1212–1218.

Arnold V.I. On the teaching of mathematics. Usp. mat. sciences. - 1998. - V. 53. - No. 1. - S. 229-234.

Ataullakhanov F.I., Zarnitsyna V.I., Kondratovich A.Yu., Lobanova E.S., Sarbash V.I. A special class of autowaves - autowaves with a stop - determines the spatial dynamics of blood coagulation. Physics – Uspekhi. - 2002. - V. 172. - No. 6. - S. 671–690.

Ataullakhanov F.I., Lobanova E.S., Morozova O.L., Shnol E.E., Ermakova E.A., Butylin A.A., Zaikin A.N. Complex modes of excitation propagation and self-organization in blood coagulation mode. Physics – Uspekhi. - 2007. - V. 177. - No. 1. - S. 87–104.

Ashcroft N., Mermin N. Solid State Physics: 2 vol. - Moscow, Mir, 1979. - Vol. 1 - 399 p .; V. 2.- 422 p.

Babkin L.B., Morozov E.M. The supercritical velocity of crack propagation in an stretched strip. Physics and Mechanics of Deformation and Fracture: Sat. MEPhI proceedings. - Moscow, Energoatomizdat, 1981. - No. 9. - S. 9–17.

Barannikova S.A. Dispersion of waves of localization of plastic deformation. Letters in ZhTF. - 2004. - V. 30. - No. 8. - S. 75–80.

Barannikova S.A., Buyakova S.N., Zuev L.B., Kulkov S.N. Deformation heterogeneity in zirconia ceramics under compression. Letters in ZhTF. - 2007. - V. 33. - No. 11. - S. 57–64.

Barannikova S.A., Nadezhkin M.V., Zuev L.B. On the localization of plastic deformation during compression of LiF crystals. Tverd. Tela.. - 2010. - V. 52. - No. 7. - S. 1291–1294.

Barannikova S.A., Nadezhkin M.V., Zuev L.B. On the relationship of the Burgers vectors of dislocations and patterns of localization of plastic deformation during compression of alkali-halide crystals. Letters in Zh. - 2011. - V. 37. - No. 16. - S. 15-21.

Barelko V.V., Barkalov I.M., Goldansky V.I., Zanin A.M., Kiryukhin D.P. Autowave mechanisms in the low-temperature chemistry of a solid. Nonlinear waves. Physics and astrophysics. - Moscow, Nauka, 1993. - S. 248–260.

Barrett C. The structure of metals. - Moscow, Metallurgizdat, 1948 . 676 p.
Bell J.F. Experimental fundamentals of the mechanics of deformable solids: in 2 vols. - Moscow, Nauka, 1984. - V. 1. - 597 p .; - V. 2. - 431 p.
Bengus V.Z. On the structure of dislocations in the crystal lattice and the mechanism of propagation of dislocations. DAN SSSR. - 1966. - V. 169. - No. 1. - S. 70–73.
Berner R., Kronmüller G. Plastic deformation of single crystals. - Moscow, Mir, 1969 .- 272 p.
Bivin Yu.K. Communication of mechanical dynamic processes and accompanying electric fields. Zh. Teor. Fiz. - 2015. - V. 85. - No. 6. - S. 69–73.
Blagoveshchensky V.V., Panin I.G. Acoustic emission during the interaction of a moving dislocation with point obstacles. Fiz. Tverd. Tela. - 2017. - V. 59. - No. 8.– S. 1554–1556.
Bobrov V.S., Lebedkin M.A. Electrical effects during low-temperature spasmodic deformation of aluminum. Tverd. Tela.. - 1989. - V. 31. - No. 6. - S. 120–126.
Bobrov V.S., Lebedkin M.A. The role of dynamic processes during low-temperature jump-like deformation of aluminum. Tverd. Tela.. - 1993. - V. 25. - No. 7. - S. 1881–1889.
Bogachev I.N., Weinstein A.A., Volkov S.D. Introduction to statistical Metallurgiya. - Moscow, Metallurgiya, 1972. - 216 p.
Boyko V.S., Natsik V.D. Elementary dislocation mechanisms of acoustic emission. Elementary processes of plastic deformation of crystals. - Kiev: Naukova Dumka, 1978. - S. 159–189.
Boyko V.S., Garber R.I., Kosevich A.M. Reversible plasticity of crystals. - Moscow, Nauka, 1991 .- 279 p.
Boyko Yu.I., Geguzin Y.E., Klinchuk Yu.I. Experimental detection of drag of dislocations by electron wind in crystals. Letters in JETP. - 1979. - V. 30. - No. 3. - S. 168–172.
Boyko Yu.F., Lubenets S.V., Ostapchuk E.I. On the structure and dynamics of dislocation ensembles arising near stress concentrators in KCl crystals. Dynamics of dislocations. - Kiev: Naukova Dumka, 1975. - S. 145–161.
Born M., Wolf E. Fundamentals of Optics. - Moscow, Nauka, 1970 .- 855 p.
Botvina L.R. Destruction. Kinetics, mechanisms, general laws. - Moscow, Nauka, 2008 .- 334 p.
Brandt N.B., Kulbachinsky V.A. Quasiparticles in condensed matter physics. - Moscow, Fizmatlit, 2007 .- 631 p.
Brown O.M., Kivshar Yu.S. The Frenkel-Kontorova model. Concepts, methods, applications. - Moscow, Fizmatlit, 2008 .- 519 p.
Bychkov R.M., Chugui Yu.V. Conversations about geometric optics. - Novosibirsk: Publishing House of the SB RAS, 2011 .- 477 p.
Weinstein A.A., Borovikov V.S. Inhomogeneity of microdeformations in a plane stress state. Probl. Prochnosti. - 1982. - No. 6. - S. 47–49.
Weinstein A.A., Kibardin M.A., Borovikov V.S. Investigation of strain heterogeneity in the AD1-M aluminum alloy. Izv. USSR Academy of Sciences. Metally - 1983. - No. 3. - S. 171–174.
Van Buren. Defects in crystals. - Moscow, IL, 1962 .- 584 p.
Vasiliev V.A., Romanovsky Yu.M., Yakhno V.G. Autowave processes. - Moscow, Nauka, 1987 .- 240 p.
West C. Holographic interferometry. - Moscow, Mir, 1982.- 504 p.
Wigner E. Studies on Symmetry. - Moscow, Mir, 1971. - 318 p.
Wirthman J., Wirthman J.R. Mechanical properties insignificantly dependent on temperature. Physical Metallurgiya. V. 3. - Moscow, Metallurgiya, 1987. - P. 112–156.
Vladimirov V.I. Collective effects in ensembles of defects. Problems of the theory of defects

in crystals. - Leningrad. Nauka, 1987. - P. 43–57.
Vladimirov V.I., Kusov A.A. Evolution of dislocation inhomogeneities during plastic deformation of metals. Fiz. Met. Metalloved. - 1975. - V. 39. - No. 6. - P. 1150–1151.
Vladimirov V.I., Romanov A.E. Disclinations in crystals. - Moscow, Leningrad, Nauka, 1986 .- 223 p.
Volterra V. The mathematical theory of the struggle for existence. - Moscow, Nauka, 1976. - 286 p.
Garofalo F. Laws of creep and long-term strength of metals. - Moscow, Metallurgiya, 1968 .- 304 p.
Gilman J.D. Microdynamic theory of plasticity. Microplasticity. - Moscow, Metallurgiya, 1972. - S. 18–37.
Gilyarov V.L., Slutsker A.I. Power engineering of a loaded quantum anharmonic oscillator. Fiz. Tverd. Tela.. - 2010a. - V. 52. - No. 3. - S. 540–544.
Gilyarov V.L., Slutsker A.I. Energy analysis of a loaded quantum anharmonic oscillator in a wide temperature range. Fiz. Tverd. Tela.. - 2010b. - V. 80. - No. 5. - S. 94–99.
Glezer A.M., Metlov L.S. Physics of megaplastic (intensive) deformation of solids. Fiz. Tverd. Tela.. - 2010. - V. 52. - No. 6. - S. 1090-1097.
Glensdorf P., Prigogine I. Thermodynamic theory of structure, stability and fluctuations. - Moscow, Mir, 1973. - 280 p.
Golubev T.M. The propagation of the front of plastic deformation in ingots during rolling at blooming. Izv. USSR Academy of Sciences. Rel. - 1950 - No. 3. - S. 401–406.
Gorbatenko V.V., Polyakov S.N., Zuev L.B. A system for visualizing plastic deformation by speckle video images. Prib. Tekh. Eksper. - 2002. - No. 3. - S. 164–165.
Gorbatenko V.V., Danilov V.I., Zuev L.B. Instability of plastic flow: Chernov – Luders bands and the Porteven – Le Chatelier effect. Zh. - 2017. - V. 87. - No. 3. - S. 372–377.
Gorelik S.S., Rastorguev L.N., Skakov Yu.A. X-ray and electron-optical analysis. - Moscow, MISiS, 1993 .- 328 p.
Granato A., Lucke K. String model of dislocation and dislocation sound absorption. Physical Acoustics. V. IVA. - Moscow, Mir, 1969. - S. 261–321.
Grigorovich V.K. The periodic law of Mendeleev and the electronic structure of metals. - Moscow, Nauka, 1966 .- 287 p.
Goodyear J., Hodge F.G. Resilience and ductility. - Moscow, IL, 1960 .- 190 p.
Davydov V.A., Zykov V.S., Mikhailov A.S. Kinematics of autowave structures in excitable media. Usp. Fiz. - 1991. - V. 161. - No. 8. - S. 45–85.
Danilov V.I., Zuev L.B. Macrolocalization of plastic deformation and staged plastic flow in polycrystalline metals and alloys. UFM. - 2008. - V. 4. - No. 4. - S. 371-422.
De Wit R. Continental theory of disclinations. - Moscow, Mir, 1977 .- 208 p.
De Groot S., Mazur P. Nonequilibrium thermodynamics. - Moscow, Mir, 1964 .- 456 p.
Davis R.M. Stress waves in solids. - Moscow, IL, 1961 .- 103 p.
Jeffries G., Swirls B. Methods of mathematical physics: in 3 vols. - Moscow, Mir, 1970.
Jones R., Wykes K. Holographic and speckle interferometry. - Moscow, Mir, 1986 .- 327 s.
Dirac P.A.M. Memories of an extraordinary era. - Moscow, Nauka, 1990 .- 207 p.
Dodd R., Aylbek J., Gibbon J., Morris H. Solitons and nonlinear wave equations. - Moscow, Mir, 1988 .- 694 p.
Dolgopolov V.T. The integer quantum Hall effect and the phenomena associated with it. Usp. Fiz. Nauk. - 2014. - V. 184. - No. 2. - S. 113–136.
Dont G. On the theory of the low-temperature maximum of internal friction in metals.. Ultrasonic methods for the study of metals. - Moscow, IL, 1963. - S. 95–118.
Dotsenko V.I., Landau A.I., Pustovalov V.V. Modern problems of low-temperature ductility of metals. - Kiev: Naukova Dumka, 1987 .- 162 p.

Dotsenko V.S. Universal randomness. Usp. Fiz. - 2011. - V. 181. - No. 3. - S. 269-292.
Drozdovsky B.A., Fridman Ya.B. The effect of cracks on the mechanical properties of structural steels. - Moscow, Metallurgizdat, 1960 .- 260 p.
Ekobori T. Physics and mechanics of fracture and solid strength. - Moscow, Metallurgiya, 1971. - 264 p.
Elenin G.G., Krylov V.V., Polezhaev A.A., Chernavsky D.S. Features of the formation of contrasting dissipative structures. DAN SSSR. - 1983. - V. 271. - No. 1. - S. 84–88.
Erokhin K.M., Kalashnikov N.P. The dependence of the binding energy of the crystal lattice of metals on the average number of conduction electrons. Fiz. Tverd. Tela.. - 2017. - V. 59. - No. 9. - S. 1667-1672.
Zhirifalko L. Statistical Solid State Physics. - Moscow, Mir, 1975 .- 382 p.
Zhurkov S.N. On the dilaton mechanism of destruction. Fiz. Tverd. Tela.. - 1983. - V. 25. - No. 10. - S. 3119–3122.
Zavodchikov S.Yu., Zuev LB, Kotrekhov V.A. Metal science issues of production of products from zirconium alloys. - Novosibirsk: Nauka, 2012 .- 255 p.
Zaikin A.I., Morozova T.Ya. Excitation propagation in an active one-dimensional medium with a trigger heterogeneity region. Biophysics. - 1978. - V. 24. - No. 1. - S. 124–128.
Zayman J. Electrons and phonons. - Moscow, IL, 1962 .- 488 p.
Zayman J. Principles of solid state theory. - Moscow, Mir, 1966 .- 416 p.
Zasimchuk E.E. Collective deformation modes, structure formation and structural instability. Cooperative deformation processes and strain localization. - Kiev: Naukova Dumka, 1989. - S. 58–100.
Zakharov V.E., Kuznetsov E.A. Solitons and collapses: two scenarios for the evolution of nonlinear wave systems. Usp. Fiz. Nauk. - 2012. - V. 182. - No. 6. - S. 570–592.
Seeger A. Sliding and hardening mechanism in cubic face-centered and hexagonal close-packed metals. Dislocations and mechanical properties of crystals. - Moscow, IL, 1960 .- S. 179–268.
Seitz F. Modern solid state theory. - Moscow and Leningrad, 1949 .- 736 p.
Zemskov E.P., Loskutov A.Yu. Oscillating traveling waves in excitable media. JETP. - 2008. - V. 134. - No. 2 (8). - S. 406-412.
Zuev L.B. Physics of electroplasticity of alkali halide crystals. - Novosibirsk: Nauka, 1990 .- 120 p.
Zuev L.B. On the formation of plasticity autowaves during deformation. MFNT. - 1994. - V. 16. - No. 10. - S. 31–36.
Zuev L.B. On the relationship between scale levels of plastic flow. MFNT. - 1996. - V. 18. - No. 5. - S. 55–59.
Zuev L.B. Steady-state autowaves of localized plasticity under the linear law of strain hardening and de Broglie relation. MFNV. - 2004. - V. 26. - No. 3. - S. 361–370.
Zuev L.B. Entropy of waves of localized plastic deformation. Letters in ZhTF. - 2005. - V. 31. - No. 3. - S. 1–4.
Zuev L.B. On the wave nature of plastic flow. Macroscopic autowaves of strain localization. Fizich. Mezomekh. - 2006a. - V. 9. - No. 3. - S. 47–54.
Zuev L.B. Autowave concept of localization of plastic deformation of solids. MFNT. - 2006b. - V. 28. - No. 9. - S. 1261–1276.
Zuev L.B. On the direct connection of lattice characteristics and parameters of localized plastic deformation. MFNT. - 2007. - v. 29. - No. 9. - S. 1147–1157.
Zuev L.B. Autowave model of plastic flow. Phys. mezomekh. - 2011. - V. 14. - No. 3. - S. 85–94.
Zuev L.B. Model of localized plasticity of solids. Autowaves and quasiparticles. MFNT. - 2012. - V. 34. - No. 2. - S. 221–238.

Zuev L.B. Crystal body as a universal generator of localized plasticity autowaves. Bulletin of the Russian Academy of Sciences. Ser. physical. - 2014. - V. 78. - No. 10. P. 1206-1213.

Zuev L.B. Macroscopic physics of plastic deformation of metals. UFM. - 2015. - V. 16. - No. 1. - S. 35-60.

Zuev L.B. Patterns of localized plastic flow as a consequence of an elastoplastic deformation invariant. MFNT. - 2016. - V. 38. - No. 10. - S. 1335–1349.

Zuev L.B. Autowave processes of localization of plastic flow in active deformable media. FMM. - 2017a. - V. 118. - No. 8. - S. 851–860.

Zuev L.B. Chernov – Luders and Porteven – Le Chatelier deformation in active deformable media of various nature. PMTF. - 2017b. - V. 58. - No. 2. - S. 164–171.

Zuev L.B., Barannikova S.A. Velocity, dispersion and entropy of autowaves for localization of plastic flow. MFNT. - 2009. - V. 31. - No. 5. - S. 711–724.

Zuev L.B., Barannikova S.A., Danilov V.I. Autowave model of plasticity of crystalline solids: macro– and microdefects. Crystallography. - 2009. - V. 54. - No. 6. - S. 1063–1073.

Zuev L.B., Barannikova S.A., Nadezhkin M.V., Gorbatenko V.V. Localization of deformation and the possibility of predicting the destruction of rocks. FTPRPI. - 2014. - No. 1. - S. 49–56.

Zuev L.B., Gromov V.E., Kurilov V.F., Gurevich L.I. The mobility of dislocations in zinc single crystals under the action of current pulses. DAN SSSR. - 1978. - V. 239. - No. 1. - S. 84–86.

Zuev L.B., Danilov V.I. Spatio-temporal ordering in the plastic flow of solids. UFM. - 2002. - V. 3. - No. 3. - S. 237–304.

Zuev L.B., Danilov V.I. Slow autowave processes during deformation of solids. Phys. mesomech. - 2003. - V. 6. - No. 1. - S. 75–94.

Zuev L.B., Danilov V.I., Barannikova S.A. Physics of macrolocalization of plastic flow. - Novosibirsk: Nauka, 2008 .- 327 p.

Zuev L.B., Danilov V.I., Gorbatenko V.V. Auto-waves of localized plastic deformation. Zh. - 1995. - V. 65. - No. 5. - S. 91–103.

Zuev L.B., Zarikovskaya N.V., Barannikova S.A., Shlyakhova G.V. Autowaves of plastic flow localization and the Hall - Petch ratio in polycrystalline Al. MFNT. - 2013. - V. 35. - No. 1. - S. 113–127.

Zuev L.B., Zarikovskaya N.V., Fedosova M.A. Macrolocalization of plastic flow in aluminum and the Hall - Petch ratio. Zh. - 2010. - V. 80. - No. 9. - S. 68–74.

Zuev L.B., Zykov I.Yu., Danilov V.I., Zavodchikov S.Yu. Inhomogeneity of the plastic flow of zirconium alloys with the parabolic law of strain hardening. PMTF. - 2000. - V. 41. - No. 5. - S. 133–138.

Zuev L.B., Semukhin B.S., Bushmeleva K.I., Zarikovskaya N.V. Ultrasound propagation velocity in Al polycrystals with different grain sizes. FMM. - 2000. - T. 89. - No. 4. - S. 111–112.

Zuev L.B., Semukhin B.S., Lunev A.G. On the relationship between the localization of plastic deformation and acoustic properties of Al. Metals. - 2004. - No. 3. - S. 99–107.

Zuev L.B., Khon Yu.A., Barannikova S.A. Dispersion of autowaves of a localized plastic flow. Zh. - 2010. - V. 80. - No. 7. - S. 53-59.

Zykov V.S. Modeling wave processes in excitable media. - Moscow, Nauka, 1984. - 165 p.

Ivanitsky G.R. Self-organizing dynamic stability of biosystems far from equilibrium. Usp. Fiz. Nauk. - 2017. - V. 187. - No. 7. - S. 757-784.

Ivanova V.S. Synergetics of fracture and mechanical properties. Synergetics and fatigue fracture of metals. - Moscow, Nauka, 1989. - S. 6–29.

Ivanova V.S., Balankin A.S., Bunin I.Zh., Oxogoev A.A. Synergetics and fractals in materials science. - Moscow, Nauka, 1994 .- 383 p.
Evens A., Rawlings R. Thermally activated deformation of crystalline materials. Thermally activated processes in crystals. - Moscow, Mir, 1973. - S. 172–206.
Imri J. Introduction to Mesoscopic Physics. - Moscow, Fizmatlit, 2002 .- 304 p.
Indenbom V.L. A dislocation description of the simplest phenomena of plastic deformation. Some problems of the physics of crystal plasticity. - Moscow, Publishing House of the Academy of Sciences of the USSR, 1960. - S. 117–158.
Indenbom V.L. The structure of real crystals. Modern crystallography. V. 2. - Moscow, Nauka, 1979. - S. 297–341.
Indenbom V.L., Orlov A.N., Estrin Yu.Z. Thermoactivation analysis of plastic deformation. Elementary processes of plastic deformation of crystals. - Kiev: Naukova Dumka, 1978. - S. 93–112.
Irwin J., Paris. P. Fundamentals of the theory of crack growth and fracture. Destruction. V. 1. - Moscow, Mir, 1976. - S. 17–66.
Kaganov M.I., Kravchenko V.Ya., Natsik V.D. Electronic braking of dislocations in metals. Usp. Fiz. Nauk. - 1973. - V. 111. - No. 4. - S. 655–682.
Kaganov M.I., Lifshits I.M. Quasiparticles. Ideas and principles of quantum solid state physics. - Moscow, Nauka, 1989 .- 96 p.
Kadich A., Edelen D. Gauge theory of dislocations and disclinations. - Moscow, Mir, 1987 .- 168 p.
Kadomtsev B.B. Dynamics and information. - Moscow, Edition of Physics-Uspekhi, 1997. - 399 p.
Kanel, G.I., Zaretsky, E.B., Razorenov, S.V., Ashitkov, S.I., and Fortov, V.E., Unusual ductility and strength of metals at ultrashort loading times, Usp. - 2017. - V. 187. - No. 5. - S. 525–545.
Kanel G.I., Fortov V.E., Razorenov S.V. Shock waves in condensed matter physics. Usp. Fiz. - 2007. - V. 177. - No. 8. - S. 809–830
Carslow G., Jaeger D. Thermal conductivity of solids. - Moscow, Nauka, 1964 .- 487 p.
Castie J. Big systems. Connectivity, complexity and disaster. - Moscow, Mir, 1982. - 216 p.
Katanaev, M.O., Geometric Theory of Defects, Usp. - 2005. - V. 175. - No. 7. - S. 705–733.
Kerkhof F. Modulation of a brittle crack by elastic waves. Physics of Fast Flowing Processes. V. 2. - Moscow, Mir, 1971. - S. 5–68.
Kerner B.S., Osipov V.V. Autosolitons. Usp. Fiz. - 1989. - V. 157. - No. 2. - S. 201–266.
Kerner B.S., Osipov V.V. Self-organization in active distributed media. Usp. Fiz. - 1990. - V. 160. - No. 9. - S. 4–73.
Kibardin M.A. The study of plastic anisotropy of a metal by a statistical method. Head. lab. - 1981. - V. 47. - No. 9. - S. 85–87.
Kibardin M.A. Development of heterogeneous distribution of plastic deformations of local regions in AD-M aluminum under different uniaxial tension conditions. - Shadrinsk: State Medical Institute, 2006. - 84 p.
Kirichenko G.I., Natsik V.D., Pustovalov V.V., Soldatov V.P., Shumilin S.E. The effect of impurities on the quantum plasticity of single crystals
β-tin. TNF. - 2010. - V. 36. - No. 4. - S. 445–451.
Kiselev, S.P., Running deformation wave in a material with strain hardening, Fizich. mesomech. - 1999. - V. 2. - No. 5. - S. 73–78.
Kiselev, S.P., Dislocation Structure of Shear Bands in Single Crystals, J. Appl. - 2006. - V. 47. - No. 6. - S. 102–113.
Kittel C. Quantum theory of solids. - Moscow, Nauka, 1967 .- 491 p.
Kittel C. Introduction to Solid State Physics. - Moscow, Nauka, 1978.- 789 p.

Klassen-Neklyudova M.V. Mechanical twinning of crystals. - Moscow, Publishing House of the Academy of Sciences of the USSR, 1960 .- 261 p.

Klimenko I.S. Holography of focused images and speckle interferometry. - Moscow, Nauka, 1985 .- 222 p.

Klimontovich Yu.L. Entropy and information of open systems. Usp. Fiz. Nauk. - 1999. - V. 169. - No. 4. - S. 443–452.

Klimontovich Yu.L. Introduction to the physics of open systems. - Moscow, Janus-K, 2002.

Klitzing, K., Quantized Hall Effect, Usp. Fiz. - 1986. - V. 150. - No. 9. - S. 107–126.

Klyavin O.V. Physics of plasticity of crystals at helium temperatures. - Moscow, Nauka, 1987 .- 255 p.

Knyazeva E.N., Kurdyumov S.P. The laws of evolution and self-organization of complex systems. - Moscow, Nauka, 1994 .- 232 p.

Kozlov A.V., Mordyuk N.S., Selitser S.I. Acoustoplastic effect with active crystal deformation. Tverd. Tela.. - 1986. - V. 28. - No. 6. - S. 1818–1823.

Kozlov E.V., Starenchenko V.A., Koneva N.A. Evolution of a dislocation substructure and thermodynamics of plastic deformation of metallic materials. Metals. - 1993. - No. 5. - S. 152–161.

Kollakot R. Diagnosis of damage. - Moscow, Mir, 1989 .- 516 p.

Kolmogorov A.N., Petrovsky I.G., Piskunov N.S. The study of the diffusion equation, coupled with an increase in the amount of substance, and its application to one biological problem. MSU Bulletin. Series A. Mathematics and Mechanics. - 1937. - V. 1. - No. 1. - S. 6–32.

Kola G. Waves of stress in solids. - Moscow, IL, 1955 .- 192 p.

Komnik S.N., Bengus V.Z. On the nature of stress relaxation in deformed crystals. DAN SSSR. - 1966. - V. 166. - No. 4. - S. 829–832.

Konovalov S.V. Regularities of the influence of weak and strong external energy influences on the plastic deformation of metals and alloys: abstract. diss. ... doctor. tech. sciences. - Novokuznetsk: Siberian state. industry Univ., 2012 .- 33 p.

Kontorova T.A., Frenkel Y.I. To the theory of plastic deformation and twinning. JETP. - 1938. - V. 8. - No. 1. - S. 89–95.

Kosevich A.M. Fundamentals of the mechanics of the crystal lattice. - Moscow, Nauka, 1972. - 280 p.

Kosevich A.M. Dislocations in the theory of elasticity. - Kiev: Naukova Dumka, 1978.- 219 p.

Kosevich A.M. Physical mechanics of real crystals. - Kiev: Naukova Dumka, 1981. - 327 p.

Kosevich A.M., Kovalev A.S. Introduction to nonlinear physical mechanics. - Kiev: Naukova Dumka, 1989 .- 300 p.

Cottrell A.H. Dislocations and plastic flow in crystals. - Moscow, Metallurgizdat, 1958.- 267 p.

Kottrel A. Theory of Dislocations - Moscow, Mir, 1969. –96 p.

Crawford F. Waves. - Moscow, Nauka, 1974.- 527 p.

Krinsky V.I., Zhabotinsky A.M. Autowave structures and prospects for their research. Autowave processes in systems with diffusion. - Gorky: IAP Academy of Sciences of the USSR, 1981. - S. 6–31.

Christian J. Theory of Transformations in Metals and Alloys. - Moscow, Mir, 1978.- 806 p.

Krishtal M.M. Interconnection of instability and mesoscopic heterogeneity of plastic deformation. FMM. - 2001. - V. 92. - No. 3. - S. 89–112.

Cracknell A., Wang K. Fermi surface. - Moscow, Atomizdat, 1978.- 350 p.

Kubo R. Thermodynamics. - Moscow, Mir, 1970 .- 304 p.

Kudrin A.B., Bakhtin V.G. Applied holography. Study of metal deformation processes. -

Moscow, Metallurgiya, 1988 .- 249 p.
Kunin I.A. Theory of dislocations. Supplement to the book:. Schouten, Y.A. Tensor analysis for physicists. - Moscow, Nauka, 1965 .- S. 371–443.
Lavrova A.I., Postnikov E.B., Romanovsky Yu.M. Brusselator - an abstract chemical reaction?. Usp. - 2009. - V. 179. - No. 12. - S. 1327–1332.
Landau L.D., Lifshits E.M. Theory of elasticity. - Moscow, Nauka, 1987 .- 248 p.
Landau L.D., Lifshits E.M. Hydrodynamics. - Moscow, Fizmatlit, 2001 .- 732 p.
Landau L.D., Lifshits E.M. Statistical Physics. V. 1. - Moscow, Fizmatlit, 2002 .- 613 p.
Landau L.D., Lifshits E.M. Quantum mechanics. Nonrelativistic theory. - Moscow, Fizmatlit, 2004 .- 797 p.
Landau L.D., Khalatnikov I.M. Viscosity Theory HeII.. JETP. - 1949. - V. 19. - No. 7. - S. 637–650.
Lapshin V.P., Lotkov A.I., Goncharova V.A., Grishkov V.N. Anisotropy of Young's modulus, shear modulus, and Poisson's ratio of the B2 phase of a Ti50Ni48Fe2 single crystal under hydrostatic compression up to 0.6 GPa. Izv. universities. Physics. - 1995. - No. 3. - S. 45–49.
Lebedev V.P. Khotkevich V.I. The effect of electric current pulses on the low-temperature (1.7–4.2 K) deformation of aluminum. FMM. - 1982. - V. 54. - No. 2. - S. 353-360.
Lebedev V.P., Krylovsky V.S. Electronic braking of dislocations in thin aluminum plates. Tverd. Tela.. - 1993. - V. 35. - No. 1. - S. 3–10.
Levich V.G. Course of theoretical physics. V. 1. - Moscow, Nauka, 1969 .- 910 p.
Lifshits I.M., Azbel M.Ya., Kaganov M.I. Electronic theory of metals. - Moscow, Nauka, 171. - 415 p.
Likhachev V.A., Malinin V.G. Structural and analytical theory of strength. - St. Petersburg: Nauka, 1993 .- 471 p.
Lobanov A.I., Kurylenko I.A., Ukrainian A.V. Autowave solutions and dissipative structures in two mathematical models of the dynamics of blood coagulation. Transactions of MIPT. - 2009. - V. 1. - No. 4. - S. 34–52.
Loskutov A.Yu., Mikhailov A.S. Introduction to Synergetics. - Moscow, Nauka, 1990 .- 270 p.
Lubenets S.V. The mobility of dislocations at low temperatures. Physical processes of plastic deformation at low temperatures. - Kiev: Naukova Dumka, 1974. - S. 220–252.
Madelung E. Mathematical apparatus of physics. - Moscow, GIFFL, 1961 .- 618 p.
Mac Lin D. Mechanical properties of metals. - Moscow, Metallurgiya, 1965 .- 431 p.
McClintock F., Argon A. Deformation and fracture of materials. - Moscow, Mir, 1970 .- 443 p.
Maksimov I.L., Sarafanov G.F., Nagorny S.N. Kinetic mechanism of slip band formation in deformable crystals. Tverd. Tela.. - 1995. - V. 37. - No. 10. - S. 3169-3179.
Malygin G.A. Kinetic mechanism of defect-free channel formation during plastic deformation of irradiated and hardened crystals. Tverd. Tela.. - 1991a. - V. 33. - No. 4. - S. 1069–1076.
Malygin G.A. Relay mechanism for the formation of dislocation-free and defect-free channels during plastic deformation of crystals. Tverd. Tela.. - 1991b. - V. 33. - No. 6. - S. 1855–1859.
Malygin G.A. The theory of the formation of cellular dislocation structures in metals. FMM. - 1991 - No. 6. - S. 31–43.
Malygin G.A. Self-organization of dislocations and slip localization in plastically deformable crystals. Tverd. Tela.. - 1995. - V. 37. - No. 1. - S. 3–42.
Malygin, G.A., Processes of self-organization of dislocations and plasticity of crystals, Usp. - 1999. - V. 169. - No. 9. - S. 979–1010.

Malygin G.A. Acoustoplastic effect and stress superposition mechanism. Tverd. Tela.. - 2000. - V. 42. - No. 1. - S. 68–75.
Malygin G.A. Analysis of strain hardening of crystals at large plastic strains. Tverd. Tela.. - 2001. - V. 43. - No. 10. - S. 1832–1838.
Malygin G.A. Analysis of the parameters of a submicron dislocation structure in metals with large plastic strains. Tverd. Tela.. - 2004. - V. 46. - No. 11. - S. 1968–1974.
Malygin, G.A., Structural Factors Influencing the Stability of Plastic Deformation during Tension of Metals with a Bcc Lattice, Fiz. - 2005a. - V. 47. - No. 5. - S. 870–875.
Malygin G.A. Analysis of structural factors determining neck formation during tension of metals and alloys with an fcc lattice. Tverd. Tela.. - 2005b. - V. 47. - No. 2. - S. 236–241.
Malygin G.A. The mechanism of strain hardening and the formation of dislocation structures in metals with large plastic strains. Tverd. Tela.. - 2006. - V. 48. - No. 4. - S. 651–657.
Malygin G.A. Strength and ductility of nanocrystalline materials and nanoscale crystals. Usp. Fiz. Nauk. - 2011. - V. 181. - No. 11. - S. 1129–1156.
Manevich L.I. Nonlinear normal modes and solitons. Appendix to the book: Scott E. Nonlinear Science. Birth and development of coherent structures. - Moscow, Fizmatlit, 2007. - S. 532–545.
Manning J. Kinetics of atomic diffusion in crystals. - Moscow, Mir, 1971. - 277 p.
Murray J. Nonlinear differential equations in biology. Lectures about models. - Moscow, Mir, 1983 .- 397 p.
Makhutov N.A. Structural strength, resource and technological safety: in 2 tons. - Novosibirsk: Nauka, 2005. - V. 1. - 493 p .; V. 2. - 610 s.
Merer H. Diffusion in solids. - Dolgoprudny: Publishing House "Intellect", 2011. - 536 p.
Miklashevich I.A. The effect of structural heterogeneity on the processes of stochastization and regularization of the processes of deformation and fracture of solids: abstract. dis. ... doctor. Phys.-Math. sciences. - Cheboksary: Chuvash state. ped Univ., 2004 .- 32 p.
Mirkin L.I. Handbook of X-ray diffraction analysis of polycrystals. - Moscow, GIFFL, 1961 .- 863 p.
Mitropolsky A.K. The technique of statistical computing. - Moscow, GIFFL, 1961 .- 479 p.
Mishchenko E.F., Sadovnichy V.A., Kolesov A.Yu., Rozov N.Kh. Autowave processes in nonlinear media with diffusion. - Moscow, Fizmatlit, 2010 .- 399 p.
Movchan B.A., Firstov S.A., Lugovskoy Yu.F. The structure, strength and fatigue resistance of microcrystalline and microlayer materials. - Kiev: Naukova Dumka, 2015 .- 170 p.
Morozov E.M. Is it possible to find the crack path immediately as a whole?. Nonlinear problems of mechanics and physics of a deformable solid. - SPb., 1998. - S. 198–212.
Morozov E.M., Polak L.S. Friedman Ya.B. On the variational principles for the development of cracks in solids. DAN SSSR. - 1966. - V. 156. - No. 3. - S. 537–540.
Morozov E.M., Smirnov Yu.I. On the possibility of describing the supercritical state of a crack. Physics and Mechanics of Deformation and Fracture: coll. MEPhI proceedings. - Moscow, Energoatomizdat, 1981. - No. 9. - S. 47-51.
Muravyov V.V., Zuev L.B., Komarov K.L. The speed of sound and the structure of steels and alloys. - Novosibirsk: Nauka, 1996 .- 183 p.
Myshlyaev M.M. Creep of polygonized structures. Imperfections in the crystalline structure and martensitic transformations. - Moscow, Nauka, 1972. - S. 194–234.
Nabarro F.R.N., Bazinsky Z.S., Holt D.B. Plasticity of pure single crystals. - Moscow, Met-

allurgiya, 1967 .- 214 p.
Nagorny S.N., Sarafanov G.F. Autowave model of the Porteven - Le Chatelier effect. Metals. - 1993. - No. 3. - S. 199–204.
Nadgorny E.M. Dynamic properties of isolated dislocations. Imperfections in the crystalline structure and martensitic transformations. - Moscow, Science, 1972. - S. 151-175.
Nazarov V.E. Dislocation nonlinearity and nonlinear wave processes in polycrystals with dislocations. Tverd. Tela.. - 2016. - V. 58. - No. 9. - S. 1665–1673.
Nye J. Physical properties of crystals. - Moscow, Mir, 1967 .- 387 p.
Naimark O.B. Nonequilibrium structural transitions as a mechanism of turbulence. Letters in ZhTF. - 1997. - V. 23. - No. 13. - S. 81–88.
Naimark O.B. Instabilities in condensed matter due to defects. Letters in JETP. - 1998. - V. 67. - No. 9. - S. 714–721.
Naimark, O.B., Collective Properties of Defect Ensembles and Some Nonlinear Problems of Plasticity and Fracture, Fizich. mesomech. - 2003. - V. 6. - No. 4. - S. 45–72.
Naimark O.B. On some regularities of scaling in ductility, fracture, and turbulence. Fizich. mesomech. - 2015. - V. 18. - No. 3. - S. 71–83.
Naimark O.B., Ladygin O.V. Nonequilibrium kinetic transitions in solids as mechanisms of localization of plastic deformation. PMTF. - 1993. - No. 3. - S. 147–154.
Nikanorov S.P., Kardashev B.K. The elasticity and dislocation inelasticity of crystals. - Moscow, Nauka, 1985 .- 253 p.
Nicolis G., Prigozhin I. Self-organization in nonequilibrium systems. - Moscow, Mir, 1979. - 512 p.
Nicolis G., Prigozhin I. Cognition of the complex. - Moscow, Mir, 1990 .- 342 p.
Novik A., Berry B. Relaxation phenomena in crystals. - Moscow, Atomizdat, 1975 .- 472 p.
Oding I.A. The theory of dislocations in metals and its application. - Moscow, Publishing House of the Academy of Sciences of the USSR, 1959. - 84 p.
Oding I.A., Ivanova V.S., Burduksky V.V., Geminov V.N. The theory of creep and long-term strength of metals. - Moscow, Metallurgizdat, 1959.- 488 p.
Olemsky A.I. Synergetics of complex systems. - Moscow, URSS, 2009 .- 379 p.
Olemsky A.I., Katsnelson A.A. Synergetics of condensed matter. - Moscow, URSS, 2003 .- 335 s.
Olemsky A.I., Petrunin V.A. Reconstruction of the condensed state of atoms under conditions of intense external action. Izv. universities. Physics. - 1987. - V. 30. - No. 1. - S. 82–121.
Olemsky A.I., Sklyar I.A. Evolution of a defect structure of a solid in the process of plastic deformation. Usp. Fiz. Nauk. - 1992. - V. 162. - No. 6. - S. 29–79.
Olemsky A.I., Kharchenko D.S. Self-organization of self-similar stochastic systems. - M .; Izhevsk: Institute for Computer Research, 2007. - 295 p.
Olemsky A.I., Khomenko A.V. Synergetics of plastic deformation. UFM. - 2001. - V. 2. - No. 2. - S. 189–263.
Orlov A.N. Introductory article. Thermally activated processes in crystals. - Moscow, Mir, 1973. - S. 5–22. \
Orlov A.N. Introduction to the theory of defects in crystals. - Moscow, Higher School, 1983. - 144 p.
Osipov K.A. Some activated processes in hard metals and alloys. - Moscow, Iz-in AN SSSR, 1962. - 131 p.
Osipov K.A. New ideas facts in metal science. - Moscow, Nauka, 1986 .- 72 p.
Ostrovsky V.S., Likhtman V.I. On the rheology of metals in surface-active media. Colloid Journal. - 1958. - V. 20. - No. 5. - S. 640–644.
Ostrovsky Yu.I., Schepinov V.P., Yakovlev V.V. Holographic interference methods for mea-

suring strain. - Moscow, Nauka, 1988 .- 209 p.
Peierls R. Quantum Theory of Solids. - Moscow, IL, 1956 .- 259 p.
Panin V.E. Fundamentals of physical mesomechanics. Fizich. mesomech. - 1998. - V. 1. - No. 1. - S. 5–22.
Panin V.E. Synergetic principles of physical mesomechanics. Fizich. mesomech. - 2000. - V. 3. - No. 6. - S. 5–36.
Panin V.E. Actual problems of physical mesomechanics of nonlinear multilevel hierarchically organized systems. Fizich. mesomechan. - 2014. - V. 17. - No. 6. - S. 5-6.
Panin V.E., Egorushkin V.E. Fundamentals of physical mesomechanics of plastic deformation and fracture of solids as non-linear hierarchically organized systems. Fizich. mesomech. - 2015. - V. 18. - No. 5. - S. 100–113.
Panin V.E., Likhachev V.A., Grinyaev Yu.V. Structural levels of deformation of solids. - Novosibirsk: Nauka, 1985 .- 229 p.
Pantel R., Puthof G. Fundamentals of quantum electronics. - Moscow, Mir, 1972. –384 p.
Patashinsky A.Z., Pokrovsky V.L. Fluctuation theory of phase transitions. - Moscow, Nauka, 1975 .- 255 p.
Pestrikov V.M., Morozov E.M. The mechanics of destruction. - St. Petersburg: Profession, 2012 .- 551 p.
Petrov Yu.V., Borodin I.N. The relaxation mechanism of plastic deformation and its justification by the example of the phenomenon of a yield tooth in whiskers. Tverd. Tela.. - 2015. - V. 57. - No. 2. - S. 336–341.
Petukhov, B.V. and Pokrovsky, V.L., Quantum and Classical Motion of Dislocations in a Peierls Potential Relief, Zh. - 1972. - V. 63. - No. 2. - S. 634–647.
Pitaevsky L.P. Macroscopic quantum phenomena. Usp. Fiz. Nauk. - 1966. - V. 90. - No. 4. - S. 623–629.
Plekhov, O.A., Naimark, O.B., SaintierN., Palin-Luc T. The elastic-plastic transition in iron: structural and thermodynamic features. Zh. - 2009. - V. 79. - No. 8. - S. 56–61.
Polak L.S., Mikhailov A.S. Self-organization in nonequilibrium physicochemical systems. - Moscow, Nauka, 1983 .- 287 p.
Polyakov S.N., Bikbaev S.A., Zuev L.B. Visualization of inhomogeneities of plastic flow by the fields of decorrelation and the speed of flicker of video speckles. Zh. - 2004. - T. 74. - No. 10. - S. 137–139.
Porubov A.V. Localization of nonlinear deformation waves. - Moscow, Fizmatlit, 2009 .- 207 p.
Presnyakov A.A. Localization of plastic deformation. - Alma-Ata: Science, 1981. - 119 p.
Presnyakov A.A., Mofa N.N. Localization of deformation of aluminum and some of its alloys under tension. Izv. USSR Academy of Sciences. Metals - 1981. - No. 2. - S. 205–208.
Prigogine I. Is the future determined? - M .; Izhevsk: Institute for Computer Research, 2005. - 240 p.
Prigogine I. From the existing to the arising. - Moscow, Nauka, 1985 .- 327 p.
Pustovalov V.V. Hopping deformation of metals and alloys at low temperatures. Phys. - 2008. - V. 34. - No. 9. - S. 871–913.
Rabotnov Yu.N. Mechanics of a deformable solid. - Moscow, Nauka, 1988 .- 712 p.
Reiser Yu.P. Physical foundations of the theory of brittle fracture cracks. Usp. Fiz. - 1970. - V. 100. - No. 2. - S. 329-347.
Rebbi K. Solitons. Physics - Uspekhi. - 1980. - T. 130. - No. 2. - S. 329–356.
Regel V.R., Slutsker A.I., Tomashevsky E.A. The kinetic nature of the strength of solids. - Moscow, Nauka, 1974. - 560 p.
Reiner M. Phenomenological macroreology. Rheology. - Moscow, IL, 1962. - S. 22–85.

Raysland J. Phonon Physics. - Moscow, Mir, 1975 .- 365 p.
Reed V.T. Dislocations in crystals. - Moscow, Metallurgizdat, 1957.- 280 p.
Roytburd A.L. Physical models of strain hardening of crystals. Physics of strain hardening of single crystals. - Kiev: Naukova Dumka, 1972. - S. 5–22.
Roytburd A.L. The theory of the formation of a heterophase structure during phase transformations in the solid state. Usp. Fiz. - 1974. - V. 113. - No. 1. - S. 69–104.
Romanovsky Yu.M., Stepanova N.V., Chernavsky D.S. Mathematical modeling in biophysics. - Moscow, Nauka, 1975 .- 343 p.
Romanovsky Yu.M., Stepanova N.V., Chernavsky D.S. Mathematical Biophysics. - Moscow, Nauka, 1984. - 304 p.
Rumer Yu.B., Ryvkin M.Sh. Thermodynamics, statistical physics and kinetics. - Novosibirsk: Publishing House of NSU, 2000 .- 608 p.
Rybin V.V. Large plastic deformation and fracture of metals. - Moscow, Metallurgiya, 1986.- 224 p.
Samarsky A.A., Zmitrenko N.V., Kurdyumov S.P., Mikhailov A.P. Thermal structures and fundamental length in a medium with nonlinear thermal conductivity and volumetric heat sources. DAN SSSR. - 1976. - V. 227. - No. 2. - S. 321–324.
Samarsky A.A., Elenin G.G., Zmitrenko N.V., Kurdyumov S.P., Mikhailov A.P. Combustion of a nonlinear medium in the form of complex structures. DAN SSSR. - 1977. - T. 237. - No. 6. - S. 1330–1333.
Sarafanov G.F. Waves of softening plastic deformation in crystals. Tverd. Tela.. - 2001. - V. 43. - No. 2. - S. 254–260.
Sarafanov G.F. Correlation effects in an ensemble of edge dislocations. Tverd. Tela.. - 2008. - V. 50. - No. 10 .- S. 1793-1799.
Sedov L.I. Continuum mechanics: in 2 vols. - Moscow, Nauka, 1970. - V. 1. - 492 p.; V. 2. - 508 s.
Selitser S.I. Random fields of internal stresses created by defects in the crystal structure. Cooperative deformation processes and localization of deformation. - Kiev: Naukova Dumka, 1989. - S. 167–195.
Shiratori M., Miyoshi T., Matsushita H. Computational fracture mechanics. - Moscow, Mir, 1986.- 334 p.
Skvortsov A.A., Litvinenko O.V. Sound radiation caused by the breakdown and stopping of edge dislocations in an isotropic medium. Tverd. Tela.. 2002. –V. 44. - No. 7. - S. 1236-1242.
Scott E. Nonlinear Science. Birth and development of coherent structures. - Moscow, Fizmatlit, 2007 .- 559 p.
Slutsker A.I. Characteristics of elementary acts in the kinetics of metal destruction. Tverd. Tela.. - 2004. - V. 46. - No. 9. - S. 1606-1613.
Slutsker A.I. Atomic level of the fluctuation mechanism of the destruction of solids (model-computer experiments). Tverd. Tela.. - 2005. - T. 47. - No. 5. - S. 777–787.
Smirnov B.I. Dislocation structure and hardening of crystals. - Moscow, Science, 1981. - 236 p.
Sobolev S.L. Transport processes and traveling waves in locally nonequilibrium systems. Usp. Fiz. - 1999. - V. 161. - No. 3. - S. 5–29.
Stepnov M.N. Probabilistic methods for assessing the characteristics of the mechanical properties of materials. - Novosibirsk: Nauka, 2005 .- 341 p.
Struzhanov V.V., Mironov V.I. Strain softening of material in structural elements. - Yekaterinburg: IMash UB RAS, 1995 .- 190 p.
Suzuki T., Yoshinaga H., Takeuchi S. Dynamics of dislocations and plasticity. - Moscow, Mir, 1989 .- 294 p.

Thomas T. Plastic flow and fracture in solids. - Moscow, Mir, 1964 .- 308 p.

Thompson, J.M.T. Instability and disaster in science and technology. - Moscow, Mir, 1985 .- 254 p.

Tretyakova T.V., Wildeman V.E. Spatial-temporal heterogeneity of the processes of inelastic deformation of metals. - Moscow, Fizmatlit, 2016 .- 120 s.

Trefilov V.I., Moiseev V.F., Pechkovsky E.P., Gornaya I.D., Vasiliev A.D. Strain hardening and fracture of polycrystalline metals. - Kiev: Naukova Dumka, 1987 .- 245 p.

Trubetskov D.I. Introduction to Synergetics. Oscillations and waves. - Moscow, URSS, 2003 .- 220 p.

Trubetskov D.I., Mchedlova E.S., Krasichkov L.V. Introduction to the theory of self-organization of open systems. - Moscow, Fizmatlit, 2002 .- 198 p.

Umezawa H., Matsumoto H., Tatiki M. Thermal field dynamics and condensed states. - Moscow, Mir, 1985 .- 504 p.

Walter G. Living Brain. - Moscow, Mir, 1966. - 300 p.

Feynman R., Leighton R., Sands M. Feynman Lectures in Physics. Vol. 3 .- Moscow, Mir, 1965 .- 238 p.

Fix V.B. Ionic conductivity in metals and semiconductors. - Moscow, Nauka, 1969 .- 295 p.

Finkel V.M. Physics of destruction. - Moscow, Metallurgiya, 1970 .- 344 p.

Franson M. Speckle Optics. - Moscow, Mir, 1980 .- 171 p.

Frenkel J.I. Introduction to metal theory. - Moscow, Nauka, 1972.- 426 p.

Frenkel J.I. Kinetic theory of liquids. - L .: Nauka, 1975 .- 592 p.

Friedel J. Dislocations. - Moscow, Mir, 1967 .- 643 p.

Friedlander F. Sound pulses. - Moscow, IL, 1962 .- 232 p.

Frittsh H. Fundamental physical constants. Usp. Fiz. - 2009. - V. 179. - No. 4. - S. 383–392.

Frost G.J., Ashby M.F. Maps of deformation mechanisms. - Chelyabinsk: Metallurgiya, 1989 .- 325 p.

Haken G. Information and self-organization. Macroscopic approach to complex systems. - Moscow, URSS, 2014 .- 317 p.

Haken G. Synergetics. Hierarchies of instabilities in self-organizing systems and devices. - Moscow, Mir, 1985 .- 419 p.

Hannanov S.H. Physical manifestations of nonlinearity in the kinetics of dislocations. FMM. - 1992. - No. 4. - S. 14–23.

Hannanov Sh.Kh., Nikanorov S.P. Collective excitations in a system of compensated dislocations. Zh. - 2007. - V. 77. - No. 1. - S. 74–78.

Hill R. Mathematical theory of plasticity. - Moscow, GITTL, 1956 .- 407 p.

Hirt J., Lot I. Theory of Dislocations. - Moscow, Atomizdat, 1972.- 599 p.

Hirsch P., Howie A., Nicholson R., Pashley D., Whelan M. Electron microscopy of thin crystals. - Moscow, Mir, 1968 .- 574 p.

Khon Yu.A., Kolobov Yu.R., Ivanov MB, Butenko A.V. The nonequilibrium state of grain boundaries and features of intrinsic grain-boundary slippage in bicrystals. Zh. - 2008. - V. 78. - No. 3. - S. 42–47.

Honeycomb R. Plastic deformation of metals. - Moscow, Mir, 1972. - 408 p.

Hudson D. Statistics for Physicists. - Moscow, Mir, 1967 .- 242 p.

Hund F. History of quantum theory. - Kiev: Naukova Dumka, 1980 .- 244 p.

Ziegler G. Extreme principles of thermodynamics of irreversible processes and continuum mechanics. - Moscow, Mir, 1966 .- 135 p.

Tsyganov M.A., Biktashev V.N., Brindley J., Holden A.V., Ivanitsky G.R. Waves in cross-diffusion systems - a special class of nonlinear waves. Usp. Fiz. Nauk. - 2007. - V. 177. - No. 3. - S. 275–300.

Chernavsky D.S. Synergetics and information. Dynamic information theory. - Moscow,

URSS, 2004 .- 287 p.
Chernavsky D.S., Starkov N.I., Scherbakov A.V. On the problems of the physical economy. Usp. Fiz. Nauk. - 2002. - V. 172. - No. 9. - S. 1045-1066.
Chernov D.K. Communication on some new observations in steel processing. D.K. Chernov and the science of metals. - Moscow, Metallurgizdat, 1950. - S. 196–207.
Chester J. Theory of Irreversible Processes. - Moscow, Nauka, 1966 .- 111 p.
Shermergor T.D. Theory of elasticity of microinhomogeneous media. - Moscow, Nauka, 1977 .- 399 p.
Shestopalov L.M. Deformation of metals and waves of plasticity in them. M .; L .: Publishing house of the Academy of Sciences of the USSR, 1958. - 268 p.
Shibkov A.A., Zolotov A.E. Nonlinear dynamics of spatio-temporal structures of macrolocalized deformation. Letters in JETP. - 2009. - V. 90. - No. 5. - S. 412-417.
Shibkov A.A., Zolotov A.E., Zheltov M.A. Mechanisms of nucleation of macrolocalized deformation bands. Izv. RAS. Ser. physical - 2012. - V. 76. - No. 1. - S. 97–107.
Shibkov A.A., Titov S.A., Zheltov M.A., Gasanov M.F., Zolotov A.E., Proskuryakov K.A., Zhigachev A.O. Electromagnetic emission during the development of macroscopically unstable plastic deformation of a metal. Tverd. Tela.. - 2016. - V. 58. - No. 1. - S. 3–10.
Schmid E., Boas V. The ductility of metals, especially metal. - Moscow, GONTI NKTP, 1938 .- 316 p.
Stremel M.A. The strength of alloys. Part 1. Defects of the lattice. - Moscow, MISiS, 1999 .- 384 p.
Stremel M.A. The strength of alloys. Part 2. Deformation. - Moscow, MISiS, 1997 .- 527 s.
Schuman V., Duba M. Analysis of deformations of opaque objects by holographic interferometry. - L.: Mechanical Engineering, 1983. - 190 p.
Ebeling B. The formation of structures in irreversible processes. - Moscow, Mir, 1979.- 279 p.
Engelke H. Theories of thermally activated processes and their application to the motion of dislocations in crystals. Thermally activated processes in crystals. - Moscow, Mir, 1973. - S. 146–171.
Ohringen A.K. Theory of micropolar elasticity. Destruction. V. 2. - Moscow, Mir, 1975. - S. 646–751.
Atkins P. Quanta. Directory of concepts. - Moscow, Mir, 1977 .- 496 p.
Eshelby J. Continental theory of dislocations. - Moscow, IL, 1963 .- 247 p.
Äbischer H.F., Waldner S. Strain distribution made visible with image-shearing speckle-interferometry. Opt. Laser Engng. - 1997. - V. 26. - N 1. - P. 407–420.
Aifantis E.C. On the microstructural origin of certain inelastic models. J. Engng. Mater. Tech. - 1984. - V. 106. - N 2. - P. 326-330.
Aifantis E.C. The physics of plastic deformation. Int. J. Plasticity. - 1987. - V. 3. - N 2. - P. 211–247.
Aifantis E.C. On the role of gradients in the localization of deformation and fracture. Int. J. Engng. Sci. - 1992. - V. 30. - N 10. - P. 1279–1299.
Aifantis E.C. Spatio-temporal instabilities in deformation and fracture. Comp. Mater. Modeling. - 1994. - V. 42. - P. 199–222.
Aifantis E.C. Pattern formation in plasticity. Int. J. Engng. Sci. - 1995. - V. 33. - N 15. - P. 2161–2178.
Aifantis E.C. Nonlinearly, periodicity and patterning in plasticity and fracture. Int. J. Non-Linear Mech. - 1996. - V. 31. - N 6. - P. 797–809.
Aifantis E.C. Gradient deformation model at nano, micro, and macro scales. J. Engng. Mater. Tech. - 1999. - V. 121. - N 1. - P. 189–202.

Aifantis E.C. Gradient plasticity. Handbook of Materials Behavior Models. - New York: Academic Press, 2001. - P. 291-307.

Abu Al-Rub R., Voyiadjis G.Z. A physically based gradient plasticity theory. Int. J. Plasticity. - 2006. - V. 22. - N 3. - P. 654-684.

Anand L., Kalidindi S.R. The process of shear band formation in plane strain com-pression of FCC metals: effect of crystallographic texture. Mech. Mater. - 1994. - V. 17. - N 1. - P. 223–243.

Argon A. Strengthening Mechanisms in Crystal Plasticity. - Oxford: University Press, 2008 .- 404 p.

Asharia A., Beaudoin A., Miller R. New perspectives in plasticity theory: dislocation nucleation, waves, and partial continuity of plastic strain rate. Math. Mech Solids.– 2008. - V. 13. - N 2. - P. 292–315.

Bass F.G., Bakanas R. Spatially localized autowave under spatio-temporal fluctuations. Waves Random Comp. Media - 2000. - V. 10. - N 2. - P. 217–229.

Bell J.F. The generation of transverse radial shear waves at a boundary between the domains of plasticity and elasticity. Int. J. Plasticity. - 1997. - V. 3. - N 1. - P. 91–114.

Billingsley J.P. The possible influence of the de Broglie momentum-wavelength relation on plastic strain "autowave" phenomena in "active materials". Int. J. Solids Structures. - 2001. - V. 38. - N 12. - P. 4221–4234.

Borg U. Strain gradient crystal plasticity effects on flow localization. Int. J. Plasticity. - 2007. - V. 23. - N 12. - P. 1400-1416.

Caillard D. Kinetics of dislocations in pure Fe. Acta Mater. - 2010. - V. 58. - N 3. - P. 3493–3515.

Caillard D., Martin J.L. Thermally Activated Mechanisms in Crystal Plasticity. - Oxford: Elsevier, 2003 .- 433 p.

Christ B.W., Picklesimer M.L. The relationship between Lüders strain, testing system compliance and other phenomenological variables affecting serrated yielding of recrystallized iron. Acta Met. - 1974. - V. 22. - N 4. - P. 435–447.

Clifton R.J. Stress wave experiments in plasticity. Int. J. Plasticity. - 1985. - V. 1. - N 4. - P. 289–302.

Coër J., Manach P.Y., Laurent H., Oliveira M.C., Menezes L.F. Piobert - Lüders plateau and Portevin - Le Chatelier effect in an Al – Mg alloy in simple shear. Mech. Res. Comm. - 2013. - V. 4. - N 1. - P. 1–7.

Counts W., Braginsky M., Battaile C., Holm E. Predicting the Hall - Petch effect in fcc metals using non-local crystal plasticity. Int. J. Plasticity. - 2008. - V. 2. - N 8. - P. 1243–1263.

Cross M.C., Hohenberg P.C. Pattern formation outside of equilibrium. Rev. Mod. Phys. - 1993. - V. 65. - N 3. - P. 851–1112.

Grimwall G., Magyari-Köpe B., Ozoliŋŝ V., Persson K.A. Lattice instabilities in metallic elements. Rev. Mod. Phys. - 2012. -V. 84. - N 2. -P. 945–986.

Danilov V.I., Narimanova G.N., Zuev L. B. On evolution of plasticity zone in the vicinity of crack tip. Int. J. Fracture. - 2000. - V. 101. - N 4. - P. L35 – L40.

Fressengeas C., Beaudoin A., Entemeyer D., Lebedkina T., Lebyodkin M., Taupin V. Dislocation transport and intermittency in the plasticity of crystalline solids. Phys. Rev. B. - 2009. - V. 79. - P. 014108-10.

Fujita H., Miyazaki S. Lüders deformation in polycrystalline iron. Acta Met. - 1978. - V. 26. - N 8. - P. 1273–1281.

Gilman J.J. Escape of dislocations from bound states by tunneling. J. Appl. Phys. - 1968. - V. 39. - N 13. - P. 6086–6090.

Gilman J.J. Microdynamics of plastic flow at constant stress. J. Appl. Phys. - 1965. - V. 36.

- N 9. - P. 2772–2777.

Gilman J.J. Micromechanics of shear banding. Mech. Mater. - 1994. - V. 17. - N 1. - P. 83–96.

Grimwall G., Magyari-Köpe B., Ozoliņŝ V., Persson K.A. Lattice instabilities in metallic elements. Rev. Mod. Phys. - 2012. - V. 84. - N 2. - P. 945–986.

Hähner P. Theory of solitary plastic waves. Appl. Phys. A. - 1995. - V. A58. - N 1. - P. 41-58.

Hallai J. F., Kyriakides S. Underlying material response for Lüders-like instabilities. Int. J. Plasticity. - 2013. - V. 47. - N 1. - P. 1–12.

Han Chin-Wu. Continuum Mechanics and Plasticity. - New York: Chapman and Hall / CRS, 2005 .- 670 p.

Hennecke T., Hähner P. Dislocation dinamics modeling of the ductile-brittle transition. Mater. Sci. Engng. - 2009. - V. 3. - P. 012005-012011.

Higashida K., Okazfki S., Takahashi T., Narita N., Morikava T., Onodera R. Crack tip dislocations and their shielding effect in MgO thin crystals. Mat. Sci. Engng. - 1997. - V. A234–236. - P. 537-540.

Higashida K., Tanaka M., Hartmaier A., Yjshino Y. Analyzing crack-tip dislocations and their shielding effect on fracture toughness. Mat. Sci. Engng. - 2008. - V. A483–484. - P. 13–18.

Higashida K., Tanaka M., Matsunaga E., Hayashi H. Crack tip stress fields revealed by infrared photoelasticity in silicon crystals. Mat. Sci. Engng. - 2004. - V. A387–389. - P. 377-380.

Horstemeyer M.F., Baskes M.I., Godfrey A., Hughes D.A. A large deformation atomistic study examining crystal orientation effect on the stress-strain relationship. Int. J. Plasticity. - 2002. - V. 18. - N 1. - P. 2023–229.

Huvier C., Conforto E., El Alami H., Delafosse D., Feaugas X. Some correlations between slip band and dislocation patterns. Mater. Sci. Engng. - 2009.– V. 3. - P. 012012–012018.

Iqbal S., Sarwar F., Rasa S.V. Quantum mechanics tunneling of dislocations: quantization and depinning from Peierls barrier. World J. Cond. Mater. Phys. - 2016. - V. 6. - N 1. - P. 103–108.

Ishii A., Li J., Ogata S. Shuffling-controlled versus strain-controlled deformation twinning: the case for HCP Mg twin nucleation. Int. J. Plasticity. - 2016. - V. 82. - N 1. - P. 32–43.

Jaoul B. Etude de la forme des courbes de deformation plastique. J. Mech. Phys. Solids. - 1957. - V. 5. - N 1. - P. 95–114.

Kobayashi M. Analysis of deformation localization based on proposed theory of ultrasonic velocity propagation in plastically deformed solids. Int. J. Plasticity. - 2010. - V. 26. - N 1. - P. 107–125.

Kovács I., Ching N.Q., Kovács-Csetènyi E. Grain size dependence of the work hardening procecc in Al99.99. phys. stat. sol. (a). - 2002. - V. 194. - N 1. - P. 3–18.

Krempl E. Relaxation behavior and modeling. Int. J. Plasticity. - 2001. - V. 17. - N 7. - P. 1419-1436.

Kubin L.P., Chihab K., Estrin Yu.Z. The rate dependence of the Portevin - Le Chatelier effect. Acta Met. - 1988. - V. 36. - N 10. - P. 2707–2718.

Kubin L.P., Estrin Yu.Z. Portevin - Le Chatelier effect in deformation with constant stress rate. Acta Met. - 1985. - V. 33. - N 3. - P. 397–407.

Kubin L.P., Estrin Yu.Z. The critical condition for jerky flow. Phys. status Sol. (b). - 1992. - V. 172. - N 1. - P. 173–185.

Kuhlmann-Wilsdorf D. The low energetic structures theory of solid plasticity. Dislocations in Solids. - Amsterdam, Boston: Elsevier, 2002. - P. 213–338.

Landau P., Shneck R.Z., Makov G., Venkert A. Evolution of dislocation patterns in fcc metals. Mater. Sci. Engng. - 2009. - V. 3. - P. 012004.
Larmor J. The influence of flaws and air cavities on the strength of materi-als. Phil. Mag. - 1892. - V. 33. - N 1. - P. 70–75.
Lebyodkin M.A., Kobelev N.P., Bougherira Y., Entemeyer D., Fressengeas C., Gornakov V.S., Lebedkina T.A., Shashkov I.V. On the similarity of plastic flow processes during smooth and jerky flow: Statistical analysis. Acta Materialia. - 2012. - V. 60. - N 22. - P. 3729–3740.
Lim H., Carroll J.D., Battaile C.C., Buchheit T.E., Boyce B.L., Wein-berger C.R. Grain-scale experimental validation of crystal plasticity finite element simulations of tantalum oligocrystals. Int. J. Plasticity. - 2014. - V. 60. - N 1. - P. 1–18.
Lüders W. Über die Äusserung der Elasticität an stahlartigen Eisenstäben and Stahlstäben und über eine beim Biegen solcher Stäbe beobachtete Molecularbewegung. Dingler's Politechhisches Jahrbuch. - 1860. - B. 155. - H. 5. - S. 18–22.
Lüthi B. Physical Acoustics in the Solids. - Berlin: Springer-Verlag, 2007 .- 418 p.
Mahajan S. Model for the FCC-HCP transformation, its applications and experimental evidence. Met. Trans. A. - 1981. - V. 12. - N 3. - P. 379–386.
Mahajan S., Green M.L., Brasen D. A model for the FCC → HCP transfor-mation, its application, and experimental evidence. Met. Trans. - 1977. - V. 8A. - N 1. - P. 283–293.
Manach P.Y., Thuillier S., Yoon J.W., Coër J., Laurent H. Kinematics of Portevin - Le Chatelier bands in simple shear. Int. J. Plasticity. - 2014. - V. 58. - N 1. - P. 66–83.
Maugin G. Sixty ears of configurationally mechanics (1950–2010). Mech. Res. Comm. - 2013. - V. 50. - N 1. - P. 39–49.
Maugin G. Solitons in elastic solids (1938–2010). Mech. Res. Comm. - 2011. - V. 38. - N 4. - P. 341–349.
McDonald R.J., Efstathiou C., Kurath P. The wave-like plastic deformation of single crystal copper. J. Engng. Mater. Technol. - 2009. - V. 131. - N 3. - P. 7–13.
Messerschmidt U.Dislocation Dynamics during Plastic Deformation. - Berlin: Springer, 2010 .- 503 p.
Meyers M.A., Nesterenko V.F., LaSalvia J.C., Qing Xue. Shear localization in dynamic deformation of materials: microstructural evolution and self-organization. Mater. Sci. Engng. - 2001. - V. A317. - N 1. - P. 204–225.
Molotskii M.I. Theoretical basis for electro– and magnetoplasticity. Mat. Sci. Engng. - 2000. - V. A287. - P. 248–258.
Mott N.F. A theory of work hardening of metal crystals. Phil. Mag. - 1952. - V. 43. - N 346. - P. 1151–1178.
Mott N.F. The mechanical properties of metals. Proc. Phys. Soc. - 1951. - V. 64. - N 5. - P. 729–741.
Mróz Z., Oliferuk W. Energy balance and identification of hardening moduli in plastic deformation process. Int. J. Plasticity. - 2002. - V. 18. - N 2. - P. 378–397.
Mudrock R.N., Lebyodkin M.A., Kurath P., Beaudoin A., Lebedkina T.A. Strain-rate fluctuations during macroscopically uniform deformation of a solid strengthened alloy. Scripta Materialia. - 2011. - V. 65. - N 6. - P. 1093–1095.
Mughrabi H. Dislocation wall and cells structures and long-range internal stresses in deformed metal crystals. Acta Metal. - 1983. - V. 31. - P. 1367–1379.
Mughrabi H. The effect of geometrically necessary dislocations on the flow stress of deformed crystals containing a heterogeneous dislocation distribution. Mat. Sci. Engng. - 2001. - V. A319–321. - P. 139–143.
Mughrabi H. On the current understanding of strain gradient plasticity. Mat. Sci. Engng. - 2004. - V. A387–389. - P. 209–213.

Mughrabi H., Pschenitzka F. Stresses to bow edge dislocation segments out of di- / multipolar edge dislocation bundles. Mat. Sci. Engng. - 2008. -
V. A483–484. - P. 469–473.
Nabarro F.R.N. Grain size, stress and creep in polycrystalline solids. Tverd. Tela.. - 2000. - T. 42. - N 8. - S. 1417-1419.
Naimark O., Davydova M. Crack initiation and crack growth as the problem of localized instability in microcrack ensemble. J. Phys. IV. - 1996. - V. 6. - N 10. - P. 259-267.
Naimark O. Structural transitions in ensembles of defects as mechanisms of failure and plastic instability under impact loading. Proc. IX Int. Conf. Fracture. Advances in Fracture Research.– 1997. - V. 6. - P. 2795–2806.
Naimark O.B. Defect induced transitions as mechanisms of plasticity and failure in multifield continua. Advances in Multifield Theories of Continua with Substructure. - Boston: Birkhauser Inc., 2003. - P. 75–114.
Nemes J.A., Eftis J. Pressure-shear waves and spall fracture described by a viscoplastic-damage constitutive model. Int. J. Plasticity. - 1992. - V. 8. - N 2. - P. 185–207.
Newnham R.E. Properties of Materials. - Oxford: University Press, 2005 .- 378 p.
Ohashi T., Kawamukai M., Zbib H. A multiscale approach for modeling scale-dependent yield stress in polycrystalline metals. Int. J. Plasticity. - 2007. –V. 23. - N 5. - P. 897–914.
Oku T., Galligan J.M. Quantum mechanical tunneling of dislocations. Phys. Rev Lett. - 1969. - V. 22. - N 12. - P. 596-597.
Oliferuk W., Maj M. Stress-strain curve and stored energy during uniaxial deformation of polycrystals. Europ. J. Mech. A / Solids. - 2009. - V. 28. - N 3. - P. 266–272.
Othmer H.G. The dynamics of forced excitable systems. Nonlinear Wave Processes in Excitable Media. - New York: Plenum Press, 1991. - P. 213–231.
Otsuka K., Shimizu K. Pseudoelasticity and shape memory effects in alloys. Int. Met. Rev. - 1986. - V. 31. - N 3. - P. 93-114.
Pelleg J. Mechanical Properties of Materials. - Dordrecht: Springer, 2013 .- 634 p.
Pontes J., Walgraef D., Aifantis E. On dislocation patterning: Multiple slip effects in the rate equation approach. Int. J. Plasticity. - 2006. - V. 22. - N 7. - P. 1486–1505.
Psakhie S.G., Zolnikov K.P., Kryzhevich D.S. Elementary atomistic processing of crystal plasticity. Phys. Let. - 2007. - V. A367. N 2. - P.250-253.
Psakhie S.G., Shilko E.V., Popov M.V., Popov V.L. The key role of elastic vortices in the initiation of intersonic shear cracks. Phys. Rev. E. - 2015 .- V.91. - P. 063302-04.
Rastogi P.K. An electronic pattern speckle shearing interferometer for the measurement of surface slope variations of three dimensional objects. Opt. Laser Engng. - 1997. - V. 26. - N 1. - P. 93–100.
Richards R. jr. Principles of Solids Mechanics. - London: CRC Press, 2001 .- 435 p.
Rizzi E., Hähner P. On the Portevin - Le Chtelier effect: theoretical modeling and numerical results. Int. J. Plasticity. - 2004. - V. 29. - N 1. - P. 121–165.
Roth A., Lebedkina T.A., Lebyodkin M.A. On the critical strain for the onset of plastic instability in an austenitic FeMnC steel. - Materials Science and Engneering. - 2012. - V. A 539. - N 1. - P. 280–284.
Schierwagen A.K. Traveling wave solutions of a nerve conduction equation for inhomogeeous axons. Non-linear Wave Processes in Excitable media. - New York: Plenum Press, 1991. - P. 107–114.
Seeger A., Frank W. Structure formation by dissipative processes in crystals with high defect densities. Non-linearPhenomenainMaterialScience. - NewYork: Trans. Tech. Publ., 1987. - P. 125–138.
Shibkov A.A., Gasanov M.F., Zheltov M.A., Zolotov A.E., Ivolgin V.I. Inter-mittent plastic-

ity associated with the spatio-temporal dynamics of deformation bands during creep tests in an AlMg polycrystal. Int. J. Plasticity. - 2016. - V. 86. - N 8. - P. 37–55.

Shibkov A.A., Denisov A.A., Zheltov M.A., Zolotov A.T., Gasanov M.T. The electric-current suppression of the Portevin - Le Shatelier effect in Al-Mg alloys. Mater. Sci. Engng. - 2014 .- V. A610. - P. 338–343.

Sjödahl M. Digital speckle photography. Digital Speckle Pattern Interferometry and Related Techniques. - New York: J. Wiley and Sons, 2001 .- P. 289–336.

Sprecher A.F., Mannan S.L., Conrad H. On the mechanisms for the electroplastic effect in metals. Acta. Met. - 1986. - V. 34. - N 7. - P. 1145–1162.

Srinivasa A.R. Large deformation plasticity and the Pointing effect. Int. J. Plasticity. - 2001. - V. 17. - N 7. - P. 1189–1214.

Steverding B. Quantization of stress waves and fracture. Mater. Sci. Engng. - 1972. - V. 9. - N 1. - P. 185–189.

Stroh A.N. The formation of crack as result of plastic flow. Proc. Roy. Soc. A. - 1954. - V. 223. - N 3. - P. 404–414.

Sun H.B., Yoshida F., Ohmori M., Ma X. Effect of strain rate on Lüders band propagation velocity and Lüders strain for annealed mild steel under unisxisl tension. Mat. Lett. - 2003. - V. 57. - N 12. - P. 4535–4539.

Tokuoka T., Iwashimizu Yu. Acoustical birefringence of ultrasonic waves in deformed isotropic elastic materials. Int. J. Sol. Struct. - 1968. - V. 4. - N 2. - P. 383-389.

Veyssiére P., Wang H., Xu D.S., Chiu Y.L. Local dislocation reaction, self-organization and hardening in single slip. Mater. Sci. Engng. 2009 / V. 3. - P. 012018-012027.

Williams R.V. Acoustic Emission. - Bristol: Adam Hilger, 1980 .- 118 p.

Wray P.J. Strain-rate of tensile failure of a polycrystalline material at elevated temperatures. J. Appl. Phys. - 1969. - V. 40. - N 10. - P. 4018–4029.

Wray P.J. Tensile plastic instability at an elevated temperature and its de-pendence upon strain rate. J. Appl. Phys. - 1970. - V. 41. - N 8. - P. 3347–3352.

Zaiser M., Aifantis E.C. Randomness and slip avalanches in gradient plasticity. Int. J. Plasticity. - 2006. - V. 22. - N 8. - P. 1432–1455.

Zaiser M., Hähner P. Oscillatory modes of plastic deformation: theoretical concept. phys. stat. sol. (b). - 1997. - V. 199. - N 2. - P. 267-330.

Zbib H.M., de la Rubia T.D. A multiscale model of plasticity. Int. J. Plasticity. - 2002. - V. 18. - N 7. - P. 1133–1163.

Zuev L.B. On a model of ductile-brittle transition be the fracture of solids. Int. J. Fracture. - 1998. - V. 90. - N 1–2. - P. L15 – L20.

Zuev L.B. Wave phenomena in low-rate plastic flow of solids. Ann. Phys. - 2001. - V. 10. - N 11–12. - P. 965–98.

Zuev L.B. The linear work hardening stage and de Broglie equation for autowaves of localized plasticity. Int. J. Solids Structures. - 2005. - V. 42. - N 3. - P. 943–949.

Zuev L.B. On the waves of plastic flow localization in pure metals and al-loys. Ann. Phys. - 2007. - V. 16. - N 4. - P. 287-310.

Zuev L.B. Autowave mechanics of plastic flow in solids. Phys. Wave Phenom. - 2012. - V. 20. - N 3. - P. 166–173.

Zuev L. B., Barannikova S. A. Evidence for the existence of localized plastic flow autowaves generated in deforming metals. Nature Sciences. - 2010a. - V. 2. - N 5. - P. 476–483.

Zuev L. B., Barannikova S. A. Plastic flow localization: autowave and quasi-particle. J. Mod. Phys. - 2010b. - V. 1. - N 1. - P. 1–8.

Zuev L. B., Danilov V.I. Plastic deformation viewed as evolution of an active medium. Int. J. Sol. Structures - 1997. - V. 34. - N 29. - P. 3795-3805.

Zuev L. B., Danilov V.I. Plastic deformation modelled as a self-excited wave process at a meso-and macro-level. Theor. Appl. Fract. Mech - 1998a. - V. 30. - N 1. - P. 175–184.

Zuev L. B., Danilov V.I. Plastic deformation viewed as a self-excited wave process. J. de Phys. IV. - 1998b. - V. 8. - P. 413-420.

Zuev L. B., Danilov V.I. A self-excited wave model of plastic deformation in solids. Phil. Mag. A. - 1999. - V. 79. - N 1. - P. 43–57.

Zuev L. B., Danilov V. I., Barannikova S. A. Pattern formation in the work hardening process of single alloyed alpha-Fe crystals. Int. J. Plasticity. - 2001. - V. 17. - N 1. - P. 47–63.

Zuev L. B., Danilov V. I., Barannikova S. A., Gorbatenko V. V. Autowave model of localized flow of solids. Phys. Wave Phenom. - 2009. - V. 17. - N 1. - P. 66–75.

Zuev L. B., Danilov V.I., Barannikova S.A., Zykov I.Yu. Plastic flow localization as a new kind of wave processes in solids. Mater. Sci. Engng. - 2001. - V. A319–321. - P. 160–163.

Zuev L. B., Danilov V.I., Kartashova N.V. Space-time self-organization of plastic deformation of FCC single crystals. JETP Let. - 1994. - V. 60. - N 7. - P. 553–555.

Zuev L. B., Danilov V. I., Kartashova N. V., Barannikova S. A. The self-excited wave nature of the instability and localization of plastic deformation. Mater. Sci. Engng. - 2001. - V. A234–236. - P. 699–702.

Zuev L. B., Danilov V. I., Poletika T. M., Barannikova S. A. Plastic deformation localization in commercial Zr-base alloys. Int. J. Plasticity. - 2004. - V. 20. - N 7. - P. 1227–1249.

Zuev L. B., Danilov V. I., Zavodchikov S. Yu., Barannikova S. A. Regular features of the evolutionary behavior exhibited by plastic flow localization and fracture in metals and alloys. J. de Phys. IV. - 1998. - V. 9. - P. 165–173.

Zuev L. B., Gorbatenko V. V., Pavlichev K.V. Elaboration of speckle photography techniques for plastic flow analysis. Measure Sci. Technol. - 2010. - V. 21. - P. 054014: 1–5.

Zuev L. B., Gromov V. E., Gurevich L. I. The effect of electric current pulses on the dislocation mobility in Zinc single crystals. phys. stat. sol. (a). - 1990. - V. 121. - N 3. - P. 437–443.

Zuev L. B., Poletika T. M. Zavodchikov S.Yu., Lositskiy A.F., Narimanova G.N., Shtutsa M.G., Belov V.I., Bocharov O.V. The self-excited waves of deformation localization and limiting state criterion for Zr-Nb alloys. J. Mater. Process. Technol. - 2002. - V. 125–126. - P. 287–294.

Zuev L. B., Semukhin B. S. Some acoustic properties of a deforming medium. Phil. Mag. A. - 2002. - V. 82. - N 6. - P. 1183–1193.

Zuev L. B., Semukhin B. S., Zarikovskaya N. V. Deformation localization and ultrasonic wave propagation rate in tensile Al as a function of grain size. Int. J. Solids Structures. - 2003. - V. 40. - N 5. - P. 941–950.

Zuev L. B., Sergeev V. P. Increase of dislocation paths in NaCl: Sr crystals by electric pulses. Phys. stat. sol. (a). - 1981. - V. 63. - N 5. - P. 119–121.

Index